Wood: Applications in Civil Engineering

Wood: Applications in Civil Engineering

Edited by
Charles Wade

MURPHY & MOORE
www.murphy-moorepublishing.com

Published by Murphy & Moore Publishing,
1 Rockefeller Plaza,
New York City, NY 10020, USA

ISBN: 978-1-63987-575-7

Cataloging-in-Publication Data

Wood : applications in civil engineering / edited by Charles Wade.
p. cm.
Includes bibliographical references and index.
ISBN 978-1-63987-575-7
1. Wood. 2. Wood--Bonding. 3. Civil engineering. I. Wade, Charles.

TA419 .W66 2022
620.12--dc23

For information on all Murphy & Moore Publications
visit our website at www.murphy-moorepublishing.com

MURPHY & MOORE

Contents

Preface VII

Chapter 1 **Wood-Boring Insect Control in Constructions by High Temperature and Microwaves** 1
Adam Krajewski

Chapter 2 **Wooden Reinforcement for Earth Constructions in the Castile Area of Spain** 20
Q. Angulo Ibáñez

Chapter 3 **Ductile Behavior of Timber Structures under Strong Dynamic Loads** 35
Maria Cristina Porcu

Chapter 4 **Wood-Reinforced Polymer Composites** 59
Anil Akdogan and Ali Serdar Vanli

Chapter 5 **Wood Thermal Properties** 81
Richard Hrčka and Marián Babiak

Chapter 6 **Traditional Wooden Buildings in China** 100
Ze-li Que, Zhe-rui Li, Xiao-lan Zhang, Zi-ye Yuan and Biao Pan

Chapter 7 **Grading of Low-Quality Wood for Use in Structural Elements** 125
Trulli Nicoletta, Monica Valdés, Barbara De Nicolo and Massimo Fragiacomo

Chapter 8 **Experimental Analyses and Numerical Models of CLT Shear Walls under Cyclic Loading** 147
Valeria Awad, Linda Giresini, Mikio Koshihara, Mario Lucio Puppio and Mauro Sassu

Chapter 9 **A Finite Element Method Model for Large Strains Analysis of Timber** 165
Vincenzo De Luca

Chapter 10 **Exterior Wood Coatings** 193
Mojgan Nejad and Paul Cooper

Chapter 11 **Flame-Retardant Systems Based on Alkoxysilanes for Wood Protection** 212
Carlos A. Giudice and Guadalupe Canosa

Permissions

List of Contributors

Index

Preface

In my initial years as a student, I used to run to the library at every possible instance to grab a book and learn something new. Books were my primary source of knowledge and I would not have come such a long way without all that I learnt from them. Thus, when I was approached to edit this book; I became understandably nostalgic. It was an absolute honor to be considered worthy of guiding the current generation as well as those to come. I put all my knowledge and hard work into making this book most beneficial for its readers.

The term wood refers to the stems and roots of trees and other woody plants. It is an organic material made of porous and fibrous structural tissues. Wood enables living trees to stand erect, and transport water and nutrients. The term can also refer to other materials engineered from wood, wood chips or fiber. Engineered wood products are used in the construction of commercial and residential buildings as structural and aesthetic materials. Even in buildings made of other materials, wood is an indispensable supporting material. A detailed explanation of the various applications of wood in civil engineering is included in this book. There has been rapid progress in this field and its applications are finding their way across multiple industries. The contents of this book will help the readers understand the modern concepts and applications of the subject.

I wish to thank my publisher for supporting me at every step. I would also like to thank all the authors who have contributed their researches in this book. I hope this book will be a valuable contribution to the progress of the field.

Editor

1

Wood-Boring Insect Control in Constructions by High Temperature and Microwaves

Adam Krajewski

Abstract

The chapter focusses on the control wood boring insects in constructions. The influence of temperature in the range of 50–65°C was studied for the larvae of *Hylotrupes bajulus* L. Attempt is made to clarify the influence of high temperature on the larvae in the context of the different results in the former German publications. The paper presents too the conditions in which the insects, which are technical pest to timber, can be combated by the use of high air temperature and microwaves 12,2cm length (2,45GHz) and which are emitted in the form of scattered field. In the experiments two prototype microwave devices ware used, their power being 600W and 1400–2400W, as well as one microwave chamber whose power was 1000 W. Laboratory results of author were presented against the designer results obtained with other insect species and the results in the recent publications by other authors. The aim of this work is to present the results of the author's microwaves research and some problems in practice. Additional experiments were conducted on polychrome and gilded timber as well as the timber into which two-inch steel nails had been driven. The study included the possibility of melting wood resin.

Keywords: insect control, old house borer, physical methods, microwaves, hot air method

1. Introduction

Because of some problems involved in the use of chemicals for insect extermination in wood, physical methods of wood-boring insect control have been applied for a long time. The insecticidal factors used in those methods include, for example, high temperature. Insects, as all living organisms, can exist only in appropriate thermic conditions. They are poikilothermic animals; however, they have a limited ability of thermic regulation by changing the intensity of breathing and evaporating. In general, however, the temperature of their bodies is preconditioned by ambient temperature.

Research on wood-boring insect control in wooden constructions was first started in Denmark [1–3]. Also in Denmark, in the 1930s, the method of exterminating the old house borer by heating wooden constructions using hot air was developed. The method was soon introduced in Germany as well. To produce masses of hot air, DEUBA heaters were applied. Hot air at the temperature of 100–120°C was inflated into attics of wooden buildings. Next, the temperature in the attics was maintained at the level of 80–100°C for several hours. Such thermic regime caused heating of the internal layers of constructional wood up to the temperature of at least 50°C. Several companies performing the old house borer extermination advertised their services in industry periodicals in the 1930s, for example, by the name of DEUBA (Hanover) and DEOB (Leipzig). After the Second World War, the method was continuingly applied on a wide scale. Specialist literature of that period is full of descriptions of the old house borer control in wooden roof constructions and the positive effects of the method [3–10].

Fundamental knowledge on thermic conditions of the old house borer extermination was acquired thanks to the research of Wichmand [1], Jensen [2], Schmidt and Schneider [11] and Becker and Loebe [12]. The results obtained in those examinations, however, were very diverse. The research was continued by Krajewski [13]. As for heating wood with solid-state objects and with infrared radiation, both methods have proven to be unpromising [13].

The phenomenon of a very limited resettlement in the wood by the old house borer after applying the heating method of insect control was also widely discussed in German periodicals at the end of the 1950s and in the early 1960s. This matter, however, of great interest and examined by the author, is outside the scope of the present chapter.

The effect of heating whole volume of wood can be also obtained using electromagnetic field of high frequencies, that is, high-frequency and ultra-high-frequency radiation, and especially microwaves. In the research, homogeneous or scattered field was applied. The term "microwaves" is usually attributed to the spectrum of electromagnetic radiation of frequencies ranging from 0.3 to 300 GHz, that is, of wavelengths ranging from 1000 to 1 mm, respectively. As far as high-frequency radio waves can be emitted in the form of both homogeneous and scattered field, in the case of microwaves, due to technical possibilities, only scattered field is available for practical purposes. According to Thomas and White [14], the first experiments on exterminating the *Lyctus brunneus* Steph. with the use of some indefinite radio waves of high frequencies were performed by J. Green in London. The research was conducted simultaneously with the experiments of MacDonald [15], concerning fighting the fungi *Serpula lacrymans* and *Coniophora puteana* with the use of microwaves of 9.1 cm wavelength. Osmolovsky [16] tested the possibilities of wood disinfestation with the use of short radio waves (frequencies ranging from 58 to 70 MHz, wavelengths ranging from 5.17 to 4.28 m). Some indefinite short radio waves were also applied by Jacquiot [17], and Wälchli and Tscholl [18] used 23-m short radio waves (13 MHz). Most research on ridding wood-boring insects from constructional wood, however, was performed using waves of frequencies ranging from 2.365 to 2.45 GHz [19–23] which correspond to wavelengths ranging from 12.7 to 12.2 cm, respectively, thus falling in the range of microwaves. After the first period of interest in applying radio waves to wood-boring insect control in the 1950s and 1960s, the subject of the use of 2.45-GHz microwaves for disinfestation was not further examined until the mid-1980s and 1990s [24–26] and the beginning of the twenty-first century [13, 27–29], with an exception for the paper by Wälchli et Tscholl [18].

Currently, a renewed interest in microwaves as a wood-boring insect control factor can be observed. Microwave heating has been recently approved by the Food and Agriculture Organization (FAO) as a significantly effective phytosanitary treatment for wood packaging materials. According to International Standards for Phytosanitary Measures (ISPM) 15 [30], target organisms are eradicated if a temperature higher than 60°C is maintained for 60 s across the entire profile of the board. The study, using pallet boards, was carried out in order to set up a treatment program that would meet ISPM 15 [30] requirements in terms of wood temperature and mortality of the insects. A 4-m long industrial tunnel oven (maximum power of 28.8 kW) was used to perform the experiments. The temperature was measured using a VarioCAM infrared camera [28, 29, 31]. The results obtained in the experiments were satisfactory. Some research in the recent years focussed only on the temperature of the wood treated with microwaves [32–34] and on the physical effects involved in such microwave processing.

In spite of a relatively large number of publications and a long-time practice in using hot air and microwaves for the aims of wood-boring insect control, the discrepancies in results obtained by researchers working independently in the field of susceptibility of well-grown old house borer larvae to heating the wood with hot air have not been explained so far. Several conditionings connected with applying microwaves to wood-boring insect control also require further explanation. It is the author's intention to provide grounds for such explanations in the present chapter, being a summary of the author's 15-year-long research on those two issues. The author's research has covered several species of wood-boring insects. However, it mainly focussed on the old house borer larvae (*Hylotrupes bajulus* L.). For this reason, the information presented here on methodology and the results of research on this particular species are more detailed. In the case of other species, the obtained results are indicated with short references only.

2. Material and methods

2.1. Experiments on the impact of high temperature on the old house borer larvae

The experiments on lethal effect of high temperature were constructed based on the results obtained by Becker and Loebe [12]. Variants of temperature and duration of action were chosen. For all experiments, well-grown individuals of the old house borer larvae (*H. bajulus* L.) were used, taken from individuals living in the wild in central and eastern Poland. The impact of high temperature was tested in the range from 50 to 65°C with 5° interspace and the following durations: temperature 50°C—duration 180, 240, 300 and 360 min; temperature 55°C—duration 20, 30, 40, 50 and 60 min; temperature 60°C—duration 5, 10 and 15 min; temperature 65°C—duration 5 and 10 min. A dryer with over doors was used. In order to prevent the possibility of temperature deviations from the temperature given in the variant of the experiment, pairs of blocks of Scots pine sapwood (*Pinus sylvestris* L.) were used, sized 15 × 25 × 50 mm³ (see **Figure 1**). One block in each pair was carved on the surface 2-mm deep. The blocks were heated for at least 2 h to the previewed temperature. The temperature of the air was measured using mercury thermometers with the accuracy of 0.5°C.

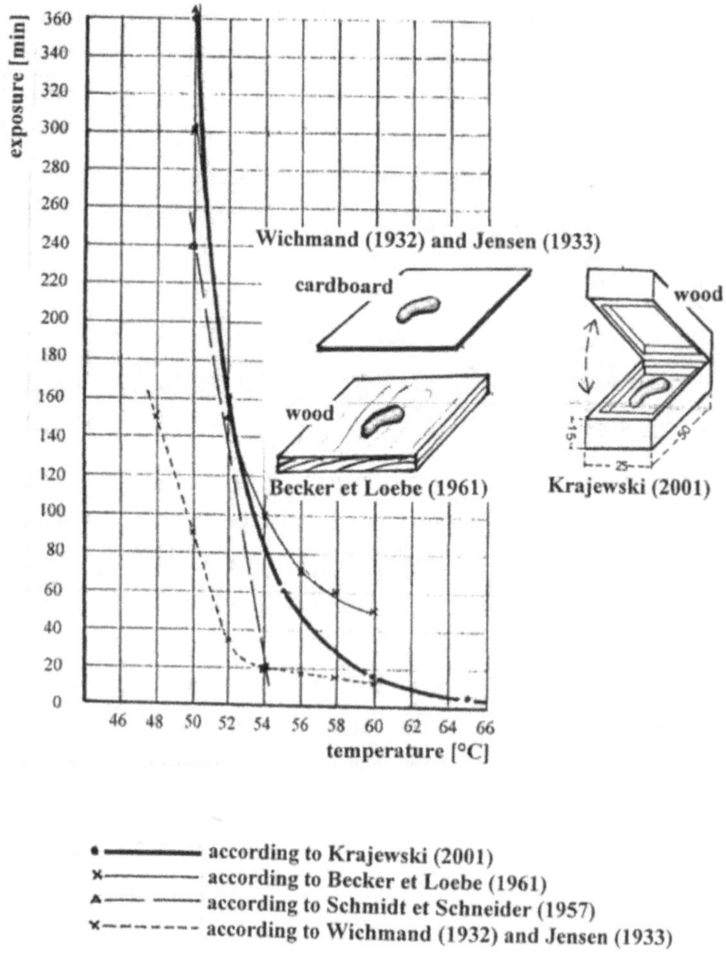

Figure 1. Effective time of exposure to high temperature in the old house borer larvae control procedures and methods of exposing larvae to hot air in different experiments (according to Krajewski [13]).

The only significant difference in relation to the experience of Becker and Loebe [12] was the way to put the larvae in the dryer. Becker and Loebe [12] installed the larvae in dryer an wooden plate (see **Figure 1**). Temperature of wood at the start was as temperature of air in the laboratory (probably about 20°C). The type of incubators in the experience of Becker and Loebe [12] was not given in the publication.

Ten old house borer larvae were used in each variant of the experiment (i.e. for each given temperature and duration), exactly as in the experiment of Becker and Loebe [12]. The masses of the old house borer larvae used in the test ranged from 20 to 230 mg (usually 70–150 mg). Each testing group of insects was compared to a control group of 10 larvae which were not exposed to high temperature. Both the testing larvae and the control group larvae were then put into wooden blocks with a glass allowing observation of behaviour of the larvae and their boring into the wood (**Figure 2**).

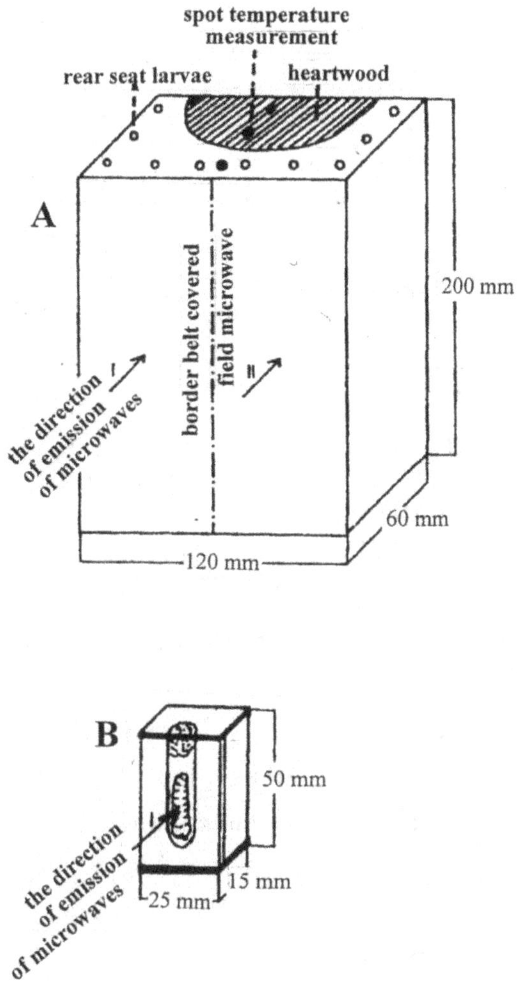

Figure 2. Samples of wood used by the author in experiments on using microwaves for the old house borer larvae control (according to Krajewski [25]). (A) Blocks used in experiments on the old house borer larvae control heartwood place of larvae introduction into the wood direction of microwave radiation. (B) Blocks used in additional observation of the larvae after microwave exposure points of temperature measurements.

2.2. Experiments on the old house borer larvae control using microwaves

Apparatus radiating microwaves of the frequency of 2.45 GHz (wavelength of 12.2 cm) in the form of scattered field was used in the experiments. Two prototypical mobile devices of 0.6 and 2.4 kW and a microwave chamber of 1 kW were produced at Wroclaw University of Science and Technology for PP PKZ Wroclaw Branch.

In the experiments, blocks sized 60 × 120 × 200 mm^3 were used (see **Figure 2**), made from:

Scots pine wood (*P. sylvestris* L.) of density ranging from 0.43 to 0.47 g/m^3 (with the average value of 0.45 g/m^3),

spruce wood of density ranging from 0.41 to 0.45 g/m^3 (with the average value of 0.43 g/m^3)

and fir wood of density ranging from 0.41 to 0.45 g/m^3 (with the average value of 0.43 g/m^3).

The radius of the heartwood semi-cylinder ranged from 30 to 40 mm. Different series of blocks were used for testing biocidal qualities of microwaves in each variant of the experiment, and for estimating the temperature corresponding to each variant.

The old house borer larvae were used in the experiments. The larvae were put into groups of 10 in each block, inside the holes drilled as shown in **Figure 2**. The length of each drilled path ranged from about 1.5 to 2 times the length of larvae body length. The diameter of each hole was a little bigger than the width of larvae. After introducing the larvae, the holes were plugged with a cotton wool cork. The blocks with larvae inside were first kept in an incubator in the temperature of 28°C, with air humidity of about 90%. After the larvae had bit into the wood, they were kept in the temperature of 20°C and air humidity of 70–75% for 12 weeks. For each variant of the experiment (i.e. power of device/time of exposure/0 mm distance between antenna and the surface of wood), three blocks were used (1 pine wood, 1 spruce wood and 1 fir wood). As a rule, microwaves were radiated by the apparatus of 0.6 kW individually onto each block, in two parallel straps of 60-mm width, onto the biggest surface opposite heartwood, with 0 mm distance between the radiating antenna and the surface of the wood. Most of the obtained results of larvae mortality and temperature measurements were handled using linear regression analysis.

Additionally, for each time of exposure, sapwood blocks sized 50 × 25 × 15 mm^3 with larvae placed individually into the carved path behind the glass (**Figure 2**) were also exposed to microwave processing. For each time variant of the experiment, 10 blocks were used (i.e. 10 larvae).

Only in a few variants of the experiment two other devices of different powers were used, as well as the 0.6 kW apparatus, with the distance between antenna and wood surface set to 100 and 200 mm. In the case of a 100- and 200-mm distance, microwaves were radiated in a single strap of 120-mm width. In those variants of experiment, single wood blocks were used.

For measuring the temperature of the wood, a platinum resistance sensor of electric current and a digital electric meter V 543 were used. Electric resistance of the sensor was measured directly after radiation inside holes drilled in the front of each block and half the length of the block, in the distance of 10, 30 and 50 mm from the surface exposed to microwave radiation. The resistance of the sensor was then converted into wood temperature. Temperature measurements were performed for the variants of the experiment with 0 mm distance between the radiating antenna and the surface of the wood (for all devices of different powers) as well as for the wood exposed to radiation from the 0.6 kW device with 100-mm distance between the antenna and the surface of the wood.

Additionally, three systems imitating chocks of constructional wood of 300 mm thickness were tested. For this purpose, three models were made from blocks of pine wood sized 60 × 120 × 200 mm^3, tightly attached by the 120 × 200 mm^2 surfaces. The 0.6 kW apparatus was used. Each model was radiated with microwaves in two straps of 60 mm width, with 0 mm distance between the antenna and the surface of the wood. Time of exposure for each model was 5, 10 and 15 min.

3. Results

3.1. Mortality of the old house borer larvae after being exposed to high temperature in the experiment with hot air

The results of the experiment on the impact of high temperature, obtained with the use of hot air, on the old house borer larvae are shown in **Table 1** and **Figure 1**. In **Figure 1** only those variants of the experiment are shown where all the larvae were killed. Results obtained in earlier experiments by other researchers are also shown in **Figure 1**.

Exposure time (min)	The average time of onset of mobility larvae after exposure (days) – in parentheses are extreme cases	The average time after exposure, when the larvae have taken a wood boring (days) – in parentheses are extreme cases	The average completion time entry larvae in the wood (days) – in parentheses are extreme cases	The level of mortality of larvae (%)
Temperature 50°C				
180	2.0	5.0 (3–6)	8.3 (6–12)	0
240	2.6 (2–4)	10.3 (7–14)	13.2 (9–17)	0
300	2.8 (2–6)	7.6 (4–12)	13.2 (10–20)	0
360	4.8 (4–6)	14.0 (12–16)	22.0 (20–24)	20
Temperature 55°C				
20	2.6 (2–4)	4.6 (4–6)	6.2 (6–7)	0
30	3.5 (3–4)	5.5 (5–6)	9.0 (8–10)	60
40	8.7 (4–16)	13.0 (6–20)	17.3 (15–23)	20
50	3.0 (2–4)	6.0 (4–8)	12.0 (10–14)	80
60	0	0	0	100
Temperature 60°C				
5	4.0 (3–5)	6.0 (7–8)	9.0 (8–10)	60
10	5.5 (4–7)	15.0 (14–16)	20.0 (19–21)	60
15	0	0	0	100
Temperature 65°C				
5	0	0	0	100
10	0	0	0	100

Table 1. The results of experiments on the effects of high temperature on old house borer larvae (hot air).

3.2. Temperature of wood heated with the use of microwaves and mortality of the old house borer larvae

Average temperatures of wood measured 10, 30 and 50 mm below the surface after heating the wood with microwaves radiated by the 0.6 kW device are shown in **Figure 3**. The process of cooling of the wood is also illustrated. The distance between the microwave radiating antenna and the surface of the wood was 0 mm in this case. In **Figure 4**, mortality rate of the old house borer larvae in pine wood, fir wood and spruce wood, determined in the experiment,

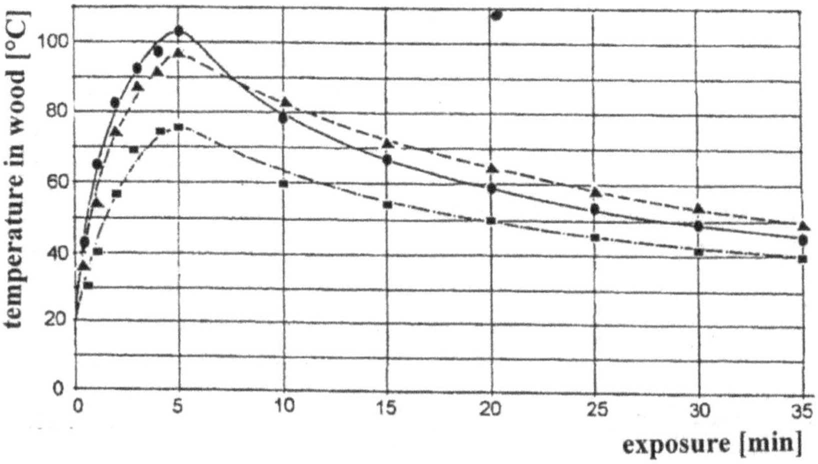

Figure 3. The process of heating in the case of wood of 60-mm thickness with microwaves (0–5 min) using the 0.6 kW device with 0 mm distance between the radiating antenna and the surface of the wood and the process of cooling of wood (5–35 min) measured 10 mm (● —), 30 mm (▲ - - - -), and 50 mm (■ – · – · – ·) below the surface (according to Krajewski [25]).

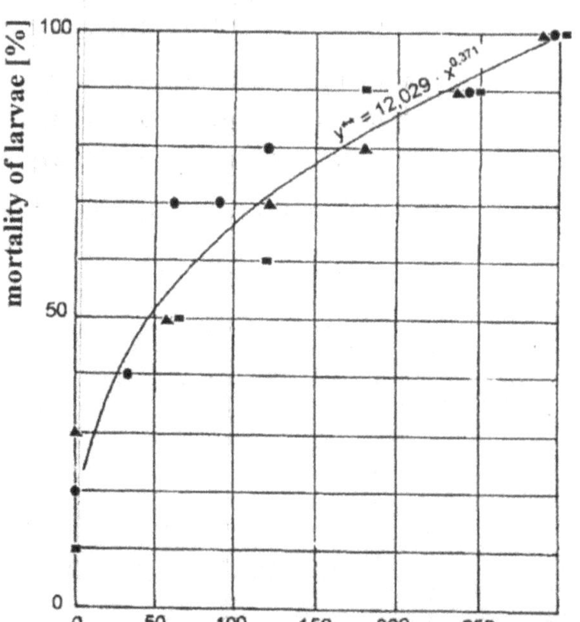

Figure 4. The relationship between mortality of the old house borer larvae and the time of exposure to microwaves radiated by the 0.6 kW device with 0 mm distance between the radiating antenna and the surface of the wood in the cases of pine wood (●), fir wood (■), and spruce wood (▲) of 60 mm thickness (according to Krajewski [25]).

is shown. In this case, the thickness of the wood was 60 mm and it was heated using microwaves produced by the 0.6 kW device with 0 mm distance between the radiating antenna and the surface of the wood.

Average temperatures of pine wood of 60 mm thickness, heated with the radiating antenna located 100 mm from the surface of the wood, are shown in **Figure 5**. Mortality rate of the old house borer larvae obtained in the experiment with 100 and 200 mm distance between the radiating antenna and the surface of the wood is shown in **Figure 6**.

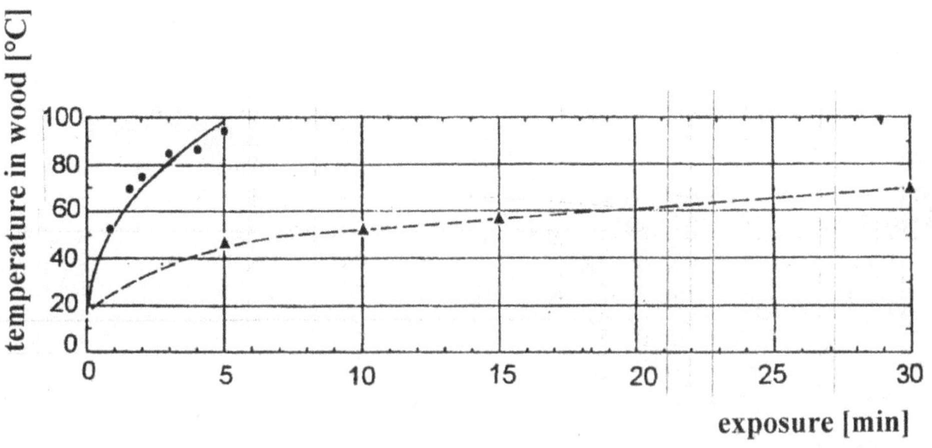

Figure 5. Average temperature of pine wood heated with microwaves radiated by the 0.6 kW device with 0 mm (● —) and 100 mm (▲ - - - -) distance between the radiating antenna and the surface of the wood (according to Krajewski [25]).

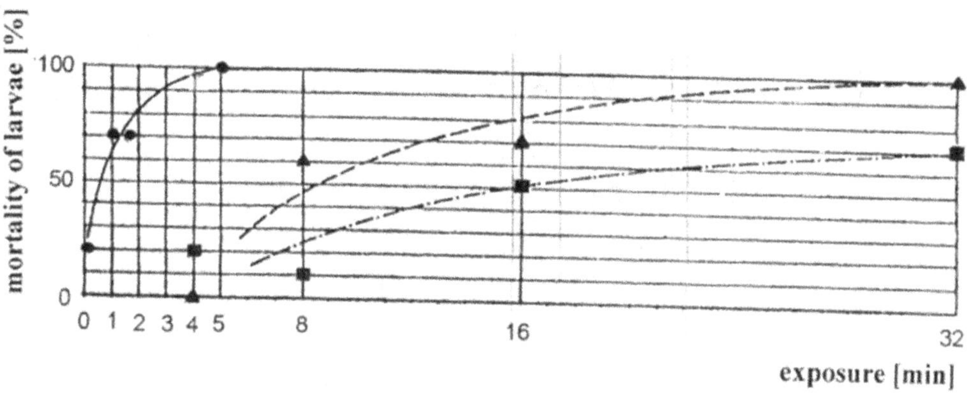

Figure 6. Mortality of the old house borer larvae after exposure to microwaves radiated by the 0.6 kW device with (● —), 100 mm (▲ - - - -), and 200 mm (■ – · – · – ·) distance between the radiating antenna and the surface of the wood (according to Krajewski [25]).

The results of heating wood of larger section (thickness of 300 mm, i.e. 5 blocks 60-mm thick, tightly attached) are shown in **Figure 7**. The graphically suggested process of thermal degradation of the wood is related to phenomena between pyrolysis and ignition 5 to 30 mm below the surface of the wood.

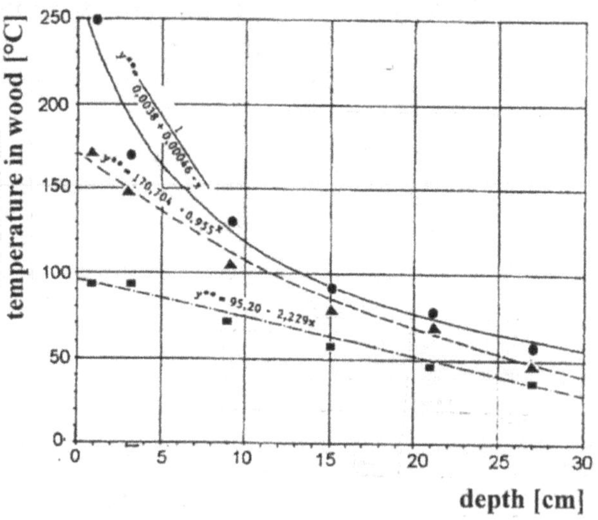

Figure 7. The temperature of different layers of wood of 300 mm thickness heated with microwaves radiated by the 0.6 kW device with 0 mm distance between the radiating antenna and the surface of the wood after 5 min (● —), 10 min (▲ - - - -), and 15 min (■ – · – · – ·) of exposure (according to Krajewski [25]).

The effect of faster heating of the wood radiated with microwaves produced by devices of higher output power is shown in **Figure 8**. In the case of 0.6 and 2.4 kW devices, the distance between the radiating antenna and the surface of the wood was 0 mm. The 1.0 kW device was a microwave chamber, refocalising the scattered microwave field.

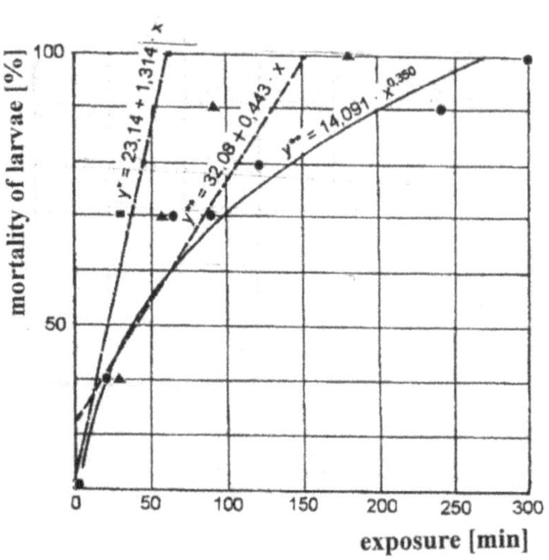

Figure 8. Mortality of the old house borer larvae after exposure to microwaves radiated by the 0.6 kW (● —), 1.0 kW (▲ - - - -), and 2.4 kW (■ – · – · – ·) devices (according to Krajewski [25]).

Unwanted effects of heating of pine wood radiated with microwaves in laboratory test conditions and in practice are shown in **Figures 9** and **10**, respectively. In **Figure 9**, pine wood blocks are shown which are used to examine the effectiveness of the old house borer larvae control.

Figure 9. Monitoring the temperature of the wood by thermal imaging can cause distortions to the view of the situation: the areas of high temperature, visible on the screen, can cover layers of much lower temperature, harmless to the larvae (photo by CARSEKT).

Figure 10. Resin melting in pine wood after exposure to microwaves radiated in practice by the 0.6 kW SAURUS device made in Italy (photo by CARSEKT).

4. Discussion

4.1. The old house borer larvae control with the use of hot air

All larvae surviving from high temperature exposure fell into thermic numbness because the duration of the procedure had not been long enough to cause mortal effects. The numbness lasted for a period of time depending on variant of the experiment (see **Table 1**). Larvae from the control group remained active for all the time, continuously carving the wood. In the experiments of Becker and Loebe [12], the phenomenon of thermic numbness of larvae was also observed if the duration of exposure to high temperature had been too short to cause death of the exposed individuals. Larvae surviving from microwave exposure also fell into numbness.

Differences in results of the relationship between larvae mortality and the temperature and duration of exposure, obtained by different researchers, result from methodological differences of their experiments. As stated in Becker and Loebe's publication [12], they placed the larvae onto a wooden plate in room temperature (about 20°C) and only after that the substrates with larvae were put inside the heating chamber. For this reason, they did not obtain the assumed homogeneous ambient temperature. Larvae were thus exposed to hot air operating from the top, while the wood in the bottom was heated only after a long period of exposure. For this reason, the results obtained by Becker and Loebe [12] are close to the author's results in the temperatures ranging from 50 to 55°C and to the results of Schmidt and Schneider [11]. Different course of the curve of relationship between mortal exposures and temperatures in the range from 54/55 and 60°C, obtained by Becker and Loebe results [12] from using wood which did not reach the temperature of the heating chamber. It seems improbable even at first glance that mortal impact of the temperature of 60°C on the old house borer larvae occurs only after as long as 50 min, unless a factor comes into play which significantly changes the assumed conditions of the experiment. In Becker and Loebe's experiment [12], only about 1/3–1/2 of the larvae body surfaces was exposed to the assumed temperature for the whole time of a given variant of the experiment. The size of the protected surface of the body depends on whether the wood was smooth or carved, a matter which is not covered by Becker and Loebe's publication [12].

A completely different location of the curve, related to the experiments of Wichmand [1] and Jensen [2], results from putting the larvae on a cardboard (again, no notice whether it was plain or corrugated) before placing them into the heating chamber. It can be assumed that the thin layer of cardboard heated much faster than wood in Becker and Loebe's experiment [12]. However, the influence of a much poorer resistance of Danish larvae (the area of Atlantic climate) as compared to the resistance of larvae from the surroundings of Berlin (much bigger impact of continental climate), suspected by Becker and Loebe [12], cannot be definitely ruled out. Larvae in Schmidt and Schneider's experiment [11] were placed inside holes drilled in previously heated wood. However, a small number of individuals (2–3) in each variant of their experiment make their results unreliable.

As for the results of the author's research [13] on resistance of various species of *Anobiidae* (*Anobium punctatum De Ger, Ptilinus pectinicornis* L., *S. paniceum* L., *E. mollis* L.) to high temperature, which are not presented here, they show that those species are much less resistant to the temperature factor than the old house borer larvae. Similar conclusions were reached by Becker and Loebe [12], who performed tests on larvae of *A. punctatum* and *L. brunneus*.

The temperature in a given place in a piece of wood can be calculated using an appropriate formula. Instead of performing time-consuming calculations, the parameters of the procedure of the old house borer control can be established in practice with the help of a suitable diagram. In order to estimate quickly the time of heating the inside of the cross section of a wooden constructional element up to the temperature of 55°C, deadly for the old house borer larvae, Schmidt et Schneider [11] presented applicable charts. The time required to heat wood to the temperature of 55°C ranges from 1 h (cross section is 100 × 140 mm^2; temperature at start time is 50°C; and temperature of inflated air is 90°C) to 14 h (cross section is 200 × 240 mm^2; temperature at start time is 0°C; and temperature of inflated air is 60°C). For example, a wooden constructional element of 180 × 180 mm^2 cross section in temperature of around 20°C at start time must be heated for 8 h if the temperature of inflated air is kept at about 60°C. According to the theory, time value taken from **Figure 1** must be added to the value of time taken from the chart. However, because of slowness of the process of heating the wood to the assumed temperature and slow returning of the heat by the wood, temperature above 40°C also acts destructively to larvae by gradual dehydration.

4.2. The old house borer larvae control using microwave radiation

The energy absorbed by a dielectric solid (Q_a), placed in a homogeneous electric field of a given intensity (E) and frequency (f), can be calculated according to the formula:

$$Q_a = 2 \cdot f \cdot \varepsilon_r \cdot \varepsilon_0 \cdot E^2 \cdot V \cdot t \cdot \mathrm{tg}\delta \quad (1)$$

where

f – frequency of the wave,

ε_r – relative permittivity of the material,

ε_0 – permittivity of free space,

E – electric intensity,

V – volume of the solid,

t – time of exposure

δ – dielectric loss angle.

Particles of water are dipoles. Placed in an electromagnetic field, for example, microwave field, they are oriented towards the rapid field changes, causing molecular friction. Apart from the qualities of the material, electric intensity and frequency of the wave are decisive. The formula given above illustrates some relationships well, yet has little practical usefulness, as modern devices, based on cavity magnetrons, produce heterogeneous (scattered) electromagnetic microwave field. Electric intensity of the field is thus varied in different points in the volume of a wooden block.

Even if the reports from research [14, 16–18] using high-frequency and ultra-high-frequency radio waves, not being microwaves, are left apart, it is still difficult to compare the results obtained by the author with the results from publications by Van den Bruel et al. [19], Bollaerts et al. [20], Berwig and Schühly [21], Watson and Moss [22], Bletchly [23], Henin et al. [28]. All those experiments, however, achieved an effective method of the old house borer larvae control, resulting from heating wood in microwave field. It is presently emphasised that for the effectiveness of the method, a temperature of 60°C must be maintained in whole interior of the wood for at least 60 s [28, 29, 31]. Wood heated with microwaves, being a good thermal insulator, allows to maintain the temperature at the level of 55–60°C for a relatively long time even after cessation of microwave radiation (see **Figure 3**). The temperature can also be maintained after cessation of electromagnetic field action. However, the temperature obtained in this way in different areas of the wood is varied. Most publications omit the effect of gas pressure, resulting from microwave radiation of the wood. It was closely noted only by Berwig and Schühly [21]. Pressure of gases, especially vapour, is so big that it causes detaching of bark from the surface of conifer wood. The effect can also act to a degree destructively to insects.

Unfortunately, scattered microwaves lose much of their effectiveness in the old house borer larvae control procedures (see **Figures 5** and **6**) which is related to the decrease of electric intensity with growing distance between the radiating antenna and the surface of the wood. The effect can be counteracted, for example, by using industrial tunnel oven, as did Henin et al. [28, 29]. A relatively big power of the applied industrial tunnel oven allows fast wood heating. The procedures of phytosanitary treatment of European pallets by microwave radiation provided for 22-mm wooden planks facilitate the task. It is also assumed that heating wood with microwave radiation concerns only wood of thickness up to 200 mm [31]. Negative effects of heterogeneous heating of wood blocks of big thickness using microwaves in the form of scattered field are shown in **Figure 7**. For this reason, application of microwave radiation to the old house borer larvae control in building constructions is a significantly more difficult task. Some unexpected thermic effects can also occur in frozen wood, as was shown by Henin et al. [31]. Also, it must be noted that monitoring the wood by thermal imaging (**Figure 9**) can cause distortions to the view of the situation. Areas illustrated with high temperatures can cover layers of wood of much lower temperatures than the temperature required to eliminate larvae. This issue concerns in particular building constructions, where there is not always the possibility to measure the temperature from two directions: from the side of microwave radiation and from the opposite side. The most heated areas of the wood are shown very bright but they can cover much colder layers. Such cases were observed in practice during the old house borer larvae extermination in wooden constructions.

Some unwanted effects of heating the wood for the aims of the old house borer larvae control include resin melting in the case of some wood species, for example, Scots pine wood. It was observed in research and was later confirmed in practical application of microwave heating in building constructions (**Figure 10**).

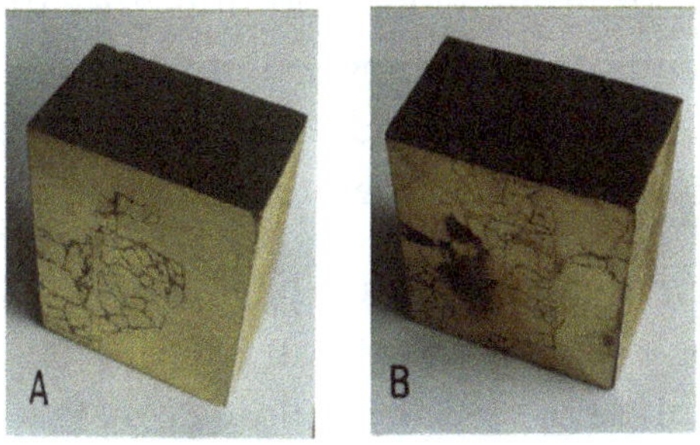

Figure 11. Condition of pine wood covered with gilding after exposure to microwaves radiated by the 0.6 kW device (photo by Krajewski): (A) 2 min and (B) 4 min.

The application of microwave heating for the aims of wood destroying insects control is limited by the presence of accompanying materials used in the construction. The author carried out ample observation, yet its results cannot be discussed in detail due to the limitations of the publication. Wood contraction, caused by desorption as a result of drying the whole volume of the wood, may result in very small cracks in both monochrome and polychrome made with some kinds of paint, for example, tempera. In the case of oil paint, binder may melt or blisters may occur on the surface of the wood. Particular threats are caused by the presence of thin electrical conductors, for example, gilding (**Figure 11**). According to Maxwell's law, electric current flows in a conductor placed in electromagnetic field. Current flow is accompanied by electrical resistance, and resistance results in heating which can even lead to ignition of the wood. The presence of 2-inch nails in the wood, exposed to microwave radiation, which guarantees effectiveness of the old house borer larvae control with a 0.6 kW device, is not dangerous. However, an attempt to destroy *Anthrenus* sp. larvae in entomological collections with entomological pins resulted in a fire.

5. Conclusions

The possibility of effective wood-boring insect control using hot air and microwave radiation was proven many years ago. However, the author's research and practical observation allow to formulate several proposals, giving a new view on some issues:

1. The differences in estimations of the old house borer larvae susceptibility to high temperatures contained in different publications result mainly from methodological differences of the performed experiments. The author's result point to purposefulness of taking up unexplained issues from pas publications, even very old ones.

2. Because of high thermic insulation of wood, practical application of wood-boring insect control both using hot air and microwaves shapes favourably in regard to theoretical indications basing on research. The biocidal effects of the suggested temperatures of 55°C

(in older publications) and 60°C (in newer publications) do not occur suddenly. Temperatures just below the theoretical limits may be hot enough to cause the effect of dehydration of larvae. In the case of microwaves, heating process is indeed very quick, but it involves releasing vapour from the wood and thus growth of pressure. Using hot air in order to heat the wood requires more time, but in this case also, besides protein denaturation, dryer effect occurs. In both methods, temperatures causing not only wood drying but also larvae dehydration are maintained for a long time.

3. It seems that microwave radiation method has better possibilities of application using industrial tunnel oven for phytosanitary treatment of wood packaging material than for wood-boring insect control in the wood of building constructions. Covering a larger space with microwave field refocalised in industrial tunnel oven and smaller thickness of wood packaging material create the possibility to obtain a more homogeneous temperature distribution in the element subjected to insects control. On the other hand, devices used to wood-boring insect control in wood of building constructions operate on the basis of "microwave brush", covering small areas of the wood. Even a small distance between the microwave radiating antenna and the surface of the wood causes huge decrease in effectiveness of the procedure.

4. Microwaves always cause some problems concerning accompanying materials in wooden constructions. Special attention must be paid to metals, including metal connectors (nails, screws etc.), also in the case of wood packaging materials. Rapid wood contraction resulting from wood drying definitely excludes applying microwaves to treatment of coated wood, even if the coating is relatively resistant to high temperatures.

Author details

Adam Krajewski

Address all correspondence to: adam_krajewski@sggw.pl

Department of Wood Sciences and Wood Protection, University of Life Sciences – Szkoła Główna Gospodarstwa Wiejskiego w Warszawie, Nowoursynowska, Warszawa

References

[1] Wichmand H. 1932: Husbukken og dens bekaempelze (Old house borer control). Teknologisk Institut Forlag, Kobnhaven.

[2] Jensen K. 1933: Wärme als Bekämpfung gegen Hausböcke (Hot as old house borer control). *Mitteilungen der Gesselschaft für Vorratschutz*, 9, 15–21.

[3] Jensen K. 1955: Jahre Hausbockbekämpfung in Dänemark (25 years of old house borer control in Denmark). *Desinfektion und Gesundheitswesen*, 47, 33–37.

[4] Keese A. 1955: Hausbockbekämpfung mit Heissluft in Deutschland (Old house borer control in Germany using hot air). *Desinfektion und Gesundheitswesen*, 47, 38–39.

[5] Weissbrodt A. 1957: Hausbockbekämpfung mit Heissluft (Old house borer control using hot air). *Desinfektion und Gesundheitswesen*, 49, 121–124.

[6] Weissbrodt A. 1958: Ein neuer Weg der Hausbockbekämpfung (New way to old house borer control). *Desinfektion und Gesundheitswesen*, 50, 79–82.

[7] Weissbrodt A., Weissbrodt W. 1959: Die weiter Entwicklung der Hausbockbekämpfung mit Heissluft (Further development of old house borer control). *Desinfektion und Gesundheitswesen*, 51, 131–134.

[8] Mathieu H. 1960: La desinsectisation des combles au moyen de lair chaud (Disinfestation of attic using hot air). *Revue de Bois*, 3, 51–53.

[9] Scholles W. 1961: Hausbockbekämpfung mit Heissluft und chemischen Mitteln (Old house borer control using hot air and chemicals). *Praktischer Schädlingsbekämpfder*, 1, 1–4.

[10] Becker G. 1962: Bekämpfung holzzerstörender Insekten mit Heissluft (Wood destroying insect control by means of hot air). *Holzshutz im Bauwesen*, 26, 19–28.

[11] Schmidt H., Schneider A. 1957: Abtötende und vorbeugende Wirkung bei der Hausbockbekämpfung mit Heissluft (Killing and preventive action in old house borer control using hot air). *Holz als Roh- und Werkstoff*, 10, 406–410.

[12] Becker G., Loebe I. 1961: Hitzempfindlichkeit holzzerstörender Käferlarven (Thermal sensitivity of wood-destroying insect larvae), *Anzeiger für Schädlingskunde*, 34(10), 145–149.

[13] Krajewski A. 2001: Fizyczne metody dezynsekcji dóbr kultury (Physical methods of wood-destroying insect control of cultural heritage), Wydawnictwo SGGW, Warszawa, 197 p.

[14] Thomas A.M., White M.G. 1959: The sterilization of insect – infested wood by high-frequency heating, *Wood*, 10, 407–410.

[15] MacDonald J.A. 1947: Experiments on the effect of centimetre-length electro-magnetic waves on wood destroying fungi, *Annals of applied Biology*, 34(3), 430–434.

[16] Osmolovsky G.E. 1950: Primienienije tokow wysokoj czastoty dlja borby z nasiekomymi – razruszitieliami driewiesiny (Use of high frequency electric current for the control of wood-destroying insects), *Truda Instituta Liesa*, 6, 162–181.

[17] Jacquiot M. 1949: L'emploi du chauffage par haute frequence pour la destruction des insects des bois mis en oeuvre (The use of heating of wood using high-frequency waves

for insect control), Comptes Rendus des Seances l'Academie d'Agriculture de France, 36(5), 637 .

[18] Wälchli O., Tscholl P. 1975: Möglichkeiten der Bekämpfung holzzerstöender Insekten ohne Giftanwendung (Possibility of wood-destroying insects control without the use of poison), *Holz als Roh- und Werkstoff*, 33(2), 49–53.

[19] Van den Bruel W.E., Bollaerts D., Pietermaat F., Stefens P. 1960: Recherches sur la destruction au moyen d'un champ electrique á tres haute frequence des insects xylophages forant le bois ouvrés (Research of very high frequency electric fields for wood-destroying insects control), *Mededelingen van de Landbouwhogeschool en de Opzoekingsstations van de Staat te Gent*, 25(3), 1377–1391.

[20] Bollaerts D., Quoilin J., Van den Bruel W.E. 1961: Nouvelles recherches relatives a l'utilisation des proprietes des micro-ondes, pour la destruction des insectes dissimules dans bois (New study of the properties of microwaves for killing wood-destroying insects). *Mededelingen van de Landbouwhogeschool en de Opzoekongsstations van de Staat te Gent*, 3, 1435–1450.

[21] Berwig W., Schühly A.F. 1964: Über Einwirkung von Mikrowellen auf im Nadelstammholz lebende Insekten (Effect of microwaves on insects living in the softwood). *Nachrichtenblatt des Deutschen Pflanzenschutzdienstes*, 2, 17–20.

[22] Watson A., Moss B.W. 1964: Microwave and moisture in building research, *British Communications and Electronics*, November, 778–781.

[23] Bletchly J.D. 1965: Very high frequency radio waves and wod-boring insect control. *Holzforschung*, 19, 47–52.

[24] Krajewski A., Zygmunt A., Burski Z. 1987: Bekämpfung der holzzerstörenden Insektenlarven mit Hilfe von Mikrovellen (Wood destroying insect larvae control using microwaves), in: Sixth International Restorer Seminar – Veszprem, Budapest 1987, 67–82.

[25] Krajewski A. 1990: Możliwość zastosowania mikrofal do dezynsekcji iglastego drewna budowlanego (Die Anwendungsmöglichkeit der Mikrowellen für die Bakämpfung der holzzerstörenden Insekten im Baunadelholz) (Possibility of the microwaves use for disinfestation of softwood in structures), Wydawnictwa PKZ (seria: studia i materiały), Warszawa, 70 p.

[26] Krajewski A. 1990: Zwalczanie ksylofagicznych owadów – szkodników technicznych drewna za pomocą mikrofal (Wood destroying insect control using microwaves), *Ochrona Zabytków*, 1, 27–34.

[27] Fleming M.R., Janowiak J.J., Kearns J., Shield J.E., Roy R., Agrawal D.K., Bauer L.S., Miller D.L., Hoover K. 2004: Parameters forscale-up of lethal microwave treatment to eradicate cerambycid larvae infested solid wood packing materials. *Forest Products Journal*, 54(7), 80–84.

[28] Henin J.-M., Leyman M., Bauduin A., Jourez B., Hébert J. 2014: Phytosanitary treatment of European pallets by microwave: developing a program to ensure compliance with ISPM 15 and monitoring its efficacy on the house longhorn beetle (*Hylotrupes bajulus* L.), *European Journal of Wood and Wood Products*. DOI 10.1007/s00107-014-0826-4, published on line 31 July 2014.

[29] Henin J.-M., Charron S., Luypaert P.J., Benoit J., Hébert J. 2014: Strategy to control the effectiveness of microwave treatment of wood in the framework of the implementation of ISPM 15, *Forest Products Journal*, 58(12), 75–81.

[30] FAO 2009: Regulation of wood packaging material in in international trade. International Standards for Phytosanitary Measures no 15 (ISPM 15), in: Food and Agriculture Organisation of the United Nations, Secretariat of the International Plant Protection Convention, Rome, Italy.

[31] Henin J.-M., Bauduin A., Leemans V., Leyman M., Benoit J., Hébert J. 2012: Microwave treatment of frozen wood packing material. Paper prepared for the joint IRG – JUFRO Research Sessions International Union of Forest Research Organizations All Division 5 Conference Estoril, Porugal, 8–13 July 2012.

[32] Zielonka P., Gierlik E. 1999: Temperature distribution during conventional and microwave wood heating. *Holz als Roh- und Werkstoff*, 57, 247–249.

[33] Rattandecho P. 2006: The simulation of microwave heating of wood using a rectangular wave guide: influence of frequency and sample size. *Chemical Engineering Science*, 61(14), 4798–4811.

[34] Hanson L., Antti A.L. 2008: Modeling microwave heating and moisture redistribution in wood. *Drying Technology*, 26,552–26,559.

2

Wooden Reinforcement for Earth Constructions in the Castile Area of Spain

Q. Angulo Ibáñez

Abstract

This chapter studies wooden reinforcements for earth constructions. Analysing vernacular houses from Castile, Spain, I discovered this reinforcement and started to compare its performance with other traditional bracings for earth construction. At present, approximately over 30% of the world's population still live in earth houses, 50% of which are in the third world countries. This is why it is so important to understand how earthwork constructions behave. Most importantly, for the rehabilitation and preservation of existing World Heritage Sites, also there is a great need to construct new buildings in developed countries under the criteria of sustainability and developing countries because of housing shortages and lack of materials. The main failure of earth constructions is because of the low tensile resistance of the earth, causing walls to detach in the corners under horizontal loads. This chapter analyses a vernacular wooden reinforcement from Castile, Spain: its history, composition, construction and structural behaviour. It compares it with earth constructions without reinforcements, using a unique model under the same conditions. This makes a qualitative and quantitative comparison possible. The conclusions can be applied to rehabilitation or construction of new-build depending on the loads, distances and height, which can be a security condition or a vital necessity.

Keywords: wooden, round pole timber, wooden bracing, vernacular construction

1. Introduction

In Spain, earth constructions have been in existence since before the arrival of the Romans [1], and the remains of an earth wall from the ninth century B.C. have been found in the substrata which dates it before the Iberians. They are more numerous in the south and central regions of the peninsula where the soil conditions are conducive to their construction. Earth wall structures were prolific across a wide spectrum of building types, from fortifications and public buildings to residential dwellings. In compression, earth as a construction material performs well structurally, but has a low tensile strength. Therefore it is important to mould, shape and condition the material to avoid tension and work in compression. In addition, the need for massively thick earthen wall sections and underperformance in their connections means that any horizontal movement (seismic activity and wind loading) could be exceptionally perilous to occupants if the risk of collapse is not considered.

Another area in the structural design of earth construction, that proves complex, is the lack of horizontal and vertical bracing between walls, floor slabs and roofs. Therefore, as floor slabs and/or roofs are not directly connected to the framework, the loads are neither distributed evenly throughout the building load, nor act to strengthen the building. Therefore, walls become independent structures under external pressures; this inherent weakness along with horizontal loading is amplified in earthquake zones. External loading conditions that cause earthen constructions to collapse, in particular adobe or rammed earth structures, manifest themselves as highlighted below [2]:

- Bending normally causes failure in the first instance. The limited tensile strength within the earth causes the walls to separate at the corners. Beginning at the top, the walls split and become individual vertical cantilevers with stability.

- Secondary failure can be due to shear forces developing in the walls. If one eliminates the above and manages the detailing between the walls, the walls are better able to withstand horizontal pressures; however, diagonal cracking may appear caused by the shear forces in the walls along the jointing.

- Finally, failure due to overturning. Once the walls fail due to bending (become independent) or fail due to shear cracking, breaking appears and they become individual vertical cantilevers. If equilibrium is not maintained, this will result in catastrophic failure and most likely total collapse. At present, earthen structures are being revised in areas prone to earthquakes. Alongside innovations in soil types and vernacular guidelines and bracing reinforcing systems [3–4], other techniques and modern elements are being incorporated, even if this means extra cost and the acquisition of materials that are not always available: columns and concrete beams as stiffeners reinforcements, attached both horizontally and vertically with earthwork infill; the integral masonry system [5]; plastering the walls with reinforced mortar or geogrid or wire mesh and cement mortar [6].

Improving soil will improve its characteristics and structural strength. Vernacularly in Spain great improvements have been made to the behaviour of adobe walls by modifying their composition (adding lime) or by bettering the constructive method (steel wall, cemented wall, etc.) [7–8]. While these innovations help to reinforce structures, they do not solve the inherent problem caused by lack of tensile strength in the connection between walls and the uppermost corners. The answer to this problem is to use one of three types of corner bracings: (i) placing ashlar blocks, quoins or rough stones into the corners of the walls; (ii) using ring beams; or (iii) embedding timber struts into the walls. These are all valid solutions to this problem.

Throughout the region of Castile, Spain, there is a vernacular wooden bracing detail employed for residential dwellings. I will endeavour to explain the method in in-depth detail and compare its resistance, structural behaviour, bracing systems, composition, materials, constructive solution and aesthetics against models where bracing is absent.

2. Description of the system

The performance analysis of the wooden reinforcement [9] is being investigated to bring an architectural, structural and construction vision, which allows the comparison between different vernacular systems of bracings for earthen structures. One option for rammed earth structures common to the Castile area is with a wooden reinforcement. It is usually commonly used in gable ends or blocks, and on sloped ground inclines to absorb tensile pressure and the resultant movements between the corner walls. While wooden reinforcements are only really seen in the Castile region, the use of stone quoins, embedding wooden struts into the uppermost section of the corners and ring beams were the most common types of bracing systems used throughout the rest of Spain and Portugal where earth was used in construction. There are three elements [10] that make up the system: the pegs, the wooden wedge, which holds them in position and the round pole reinforcement itself. These hold the walls together using the wedges and pegs on the exterior to limit lateral movement. This was conceived as a way of absorbing tensile forces in the upper corners of the wall's joint where there is an absence of beams, and utilizing the compressive force inherent in the round pole reinforcement and the pegs (**Figure 1**).

The reinforcement is usually a straight de-barked round pole timber, (also used to create beams, tie beams, lintels etc.) with a radius of approximately 10 and 25 cm (depending on the house type). At alternate ends a hole is drilled to house the pegs; the hole is situated in a proximity of 30–40 cm from the extremities to prevent the pole from splitting. The pegs or wooden pegs can be straight or curved (made from squared-off trunks or sawn-off planks or metallic pegs, or plough tails). They are positioned in the holes, parallel or perpendicular to the floor construction for build ability and hammered into position. During construction wedges are used to jam the pegs in place, either with one wedge, or with two (one on top of the other crossing over at the points). Throughout the lifespan of the building, either due to wear and tear or weathering, the wooden pole may warp, and problems may be avoided by

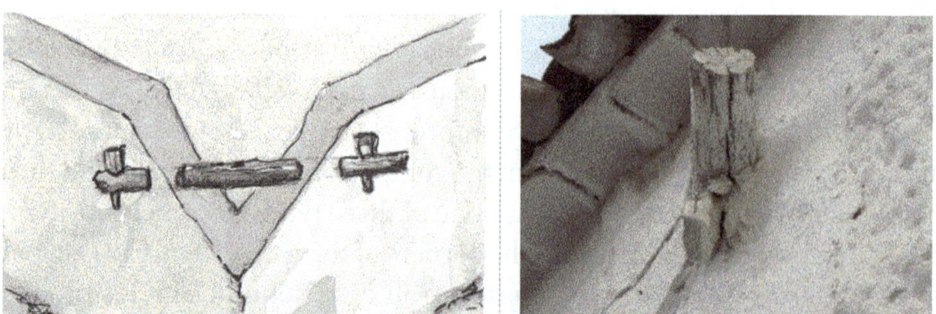

Figure 1. Components of the wooden reinforcement system: Earth wall—reinforcement—peg—wedge.

adjusting the pegs against the walls and by adding new wedges. Larger load-bearing wedges are usually used to fix the pegs to the half lap joint hooks. The wood used in the Castile area is Aleppo pine and the common pine tree. It is also possible that stronger woods are used for the pegs and wedges, such as holm oak and common oak.

The sizes differ vastly. For larger spans, a longer round pole with a greater diameter is employed. Analysing the evolution of this system, builders who built with rammed earth were well aware of the mechanics of compression bracing and its design. The loadings placed on the soil during the construction phase of rammed earth walls over timbers were counteracted by wooden cross members, iron struts or esparto fixings to create the shuttering to hold the soil. Consequently, the transferability of one technology to the other was possible. Wooden struts embedded into the corners and/or ridge beams for bracing systems were found in some of the oldest and most important earth constructions in the area. In some cases, additional, diagonal structural supports were employed working together with the corner bracings described above. From another standpoint, in the Castile region builders were limited to wooden reinforcements for vernacular dwellings. This method shows a more simplistic approach than embedded struts or ridge beams. In poor vernacular farmhouses, wooden reinforcement fulfilled all needs and is an economical, ingenious and simple solution using a minimum amount of materials and labour. Furthermore, as they are visible, they can be easily maintained and inspected.

The wooden bracings are simple wall connectors, looking from a structural and construction viewpoint [9] (both parallel and perpendicular), absorbing tensile loading at the top of the corner joints. Being exposed to the exterior, permitting ease of access for maintenance purposes and, in addition, the pegs can be simply readjusted or replaced when subjected to wear and tear. One disadvantage is that without suitable protection at the ends of the round poles where the pegs are placed, wood rot may take hold and result in modifications needed or even total system replacement. Another issue is aesthetical; with the cantilevering of the extremities, the pegs and wedges are in full sight on the façade and in addition the diagonal member obstructs the full use of the internal corner.

Maintenance and general upkeep in earth soil system construction, good detailing and workmanship are essential to prevent water ingress into the structure. Another extremely necessary element to maintain is the rendering on the façade. This acts as a rain screen for the earthen walls and keeps the walls and timber bracing structure dry to prevent failure.

The lack of a protective coating on the exterior wood may cause a knock-on effect weakening the timber bracing, which in turn could cause failure in the wall itself. The protection of the beams normally consists of three techniques: the initial wall covering (if the cantilever of the bracing structure was minimal, the earth-lime wash would simply be employed over it); with terracotta pan tiles embedded over the timber extremities and also protecting the pegs and wedges; and in recent years zinc panelling nailed directly on top of the wood. It is worth noting again that having the bracings visible makes inspecting their state, and, if necessary, making any adjustments or substitutions much easier in relation to general maintenance. It is not uncommon that as time goes by, the variations in temperature and humidity warp the wood, making it necessary to replace pegs or new wedges.

3. Calculation techniques. Description of the models, materials and loads

The calculation uses a scalar damage model for frictional plastic materials, with an application developed by the Polytechnic University of Valencia. In the CID (Calculation & Structural Design & Foundations), structural analysis program for CAD environments building structures [11, 12], a programme has been initiated referring to the isotropic damage model developed in the last 20 years. This programme is based on damage mechanics, which is part of the internal variables that introduce microstructural changes in the behaviour of materials, modelling the influence of the history of material behaviour in the evolution of stresses. With the correct description of the damage function representing the material response in compression and tension, you can model the non-linear performance of the soil using the damage theory. The appearance of fissures and their progression over time describe trajectories of numerous damaged spots, represented as an effect of local damage in terms of material parameters and functions that control the progression of damage to the successive state of tension at each point. This programme has been calibrated with several examples and studies as well as with existing physical elements [13, 20].

The typological model is a vernacular house with two floors above ground of 7.20 x 9.20 m (façade x dividing wall) and load-bearing wall parallel to façade for supporting floor slab and ridge beam. Load-bearing walls are of rammed earth, 40–60 cm in section depending on their slenderness and loads. The floor construction is made up of timber struts 15 cm in diameter at 50-cm centres, filled with supporting vaults (mud bricks and loam) or wattle and mortar on top of the beams. Pitched roof made with round pole timbers, wattle and clay supports for tiles on the load-bearing walls (façade and intermediate wall). Ground height of 3.90 m and 6.00 m ridge. A ground floor of height 2.5 m. The structural model is discretized with finite hexahedra solid elements (volumetric) for earthen walls and finite bar elements (linear) in order to replace beams and reinforcements supported at the solid nodes and substituting floor infill for the appropriate loads. The model has 1.972 hexahedra of 0.20 x 0.40 x 0.40 m per side with 8 nodes each, 61 bars for roof and slab beams and 9 bars for lintels.

In attempts to analyse the influence of the wooden reinforcement in earth constructions, the variants of the walls in terms of their composition (single, hooped, linked, reinforced with lime…) or the composition of materials used (earth, improved, gravel, ceramic pieces…) haven't been included [7–8]. Basic physic-mechanics and the general qualities of the soil have been employed, without improving the composition of the construction, materials or treatments, applying them to models of wooden braced and non-braced construction so that they are equivalent and consequently achieving a common result, without depending on the traditions of each site or of each skill that the tradesmen employed.

Evidently, an improvement in the material or the composition of the walls generally implies an improvement in the structural behaviour of this combination. Soil characteristics of the corner elements were defined with less mechanical resistance because of the difficulty of creating the corners inside the frameworks and/or poor joints with vertical recess solutions. Middle and conservative physic-mechanical properties have been adopted for materials from the results of tests (from the Castile area) and literature [3, 7, 21, 22] (**Table 1**).

	Deformation E (N/mm^2)	Poisson	Density (Kg/m^3)	Compressive Resistance Fc (N/mm^2)	Shear Resistance Ft (N/mm^2)
Earth	500	0,2	2.000	1	0,025
Earth of the corners	500	0,2	2.000	60	4,5
Wood	11.000	0,25	550	12	10

Table 1. Physic-mechanics characteristics of materials used.

For the hypothesis of loads and load combinations we have adopted the values of official documents and regulations:

- Dead loads: values from the tests results.

- Live loads: based on current Spanish legislation [21].

- Earthquakes: according to the Spanish legislation [22]. Values have been taken to analyse the worst-case scenario, although this regulation would preclude the construction of soil-based buildings under such conditions.

In the analysis three methods were used:

- Linear static calculation: based on the hypothesis of linear elastic performance of materials and noting the balance of the structure without becoming deformed. Loads and load combinations are considered for the two main directions.

- Non-linear static calculation: this highlights the stress-strain performance of non-linear material and geometric non-linearity, i.e. achieving balance of the structure in its deformed state. We analysed four independent load combinations for the two main directions, introducing proportional increases in 20 steps, taking into account geometric variations and materials:

- Gravitational loads (dead and live loads) without majority.

- Gravitational loads (dead and live loads) and horizontal (wind) without majority.

- Gravitational loads (dead and live loads) to collapse.

- Gravitational loads (dead and live loads) and horizontal (wind) to collapse.

- Dynamic-seismic calculation: we have analysed two equivalent static load combinations for earthquakes for the two main directions of the model.

The current study has concentrated on the comparison between un-reinforcement mud walls and those with wooden reinforcement:

Model 1.- Earth walls without reinforcements.

Rammed earth walls with corner framework or making a vertical recess in the finished wall so when the two walls are put together they join perfectly. 1A. -Earth walls 40-cm thick without bracings. This is the base model for implementing the analysed bracings and comparing performances and results. It is employed as the reference. 1B. -Earth walls 60-cm thick without bracings. This model tests the influence of the thickness on the structure performance against the loads.

Model 2.- Rammed earth walls 40-cm thick with wooden reinforcements in the corners with struts of 15 cm in diameter, 1 m from the interior corner. Models where the wooden reinforcement has been applied to the four superior corners of the first floor with rammed earth walls 40-cm thick.

4. Experimental

Analysing the efforts obtained either from a load combination or the whole load combination, we are able to measure the performance of the structure and see the areas where the force exceeds the material's point of resistance.

- Linear static method under gravity loads and wind.

In the model without bracing we can see that the major pressures are felt in the upper joints between the walls. However, the model shows less pressure on the joints between walls; it is better distributed towards the reinforcement joint (**Figure 2**). In **Figure 3** we can see in detail the superior wall joints. The pressure produced on the corners is greater on the non-reinforced model and in the wooden reinforcement model we can see that this pressure is

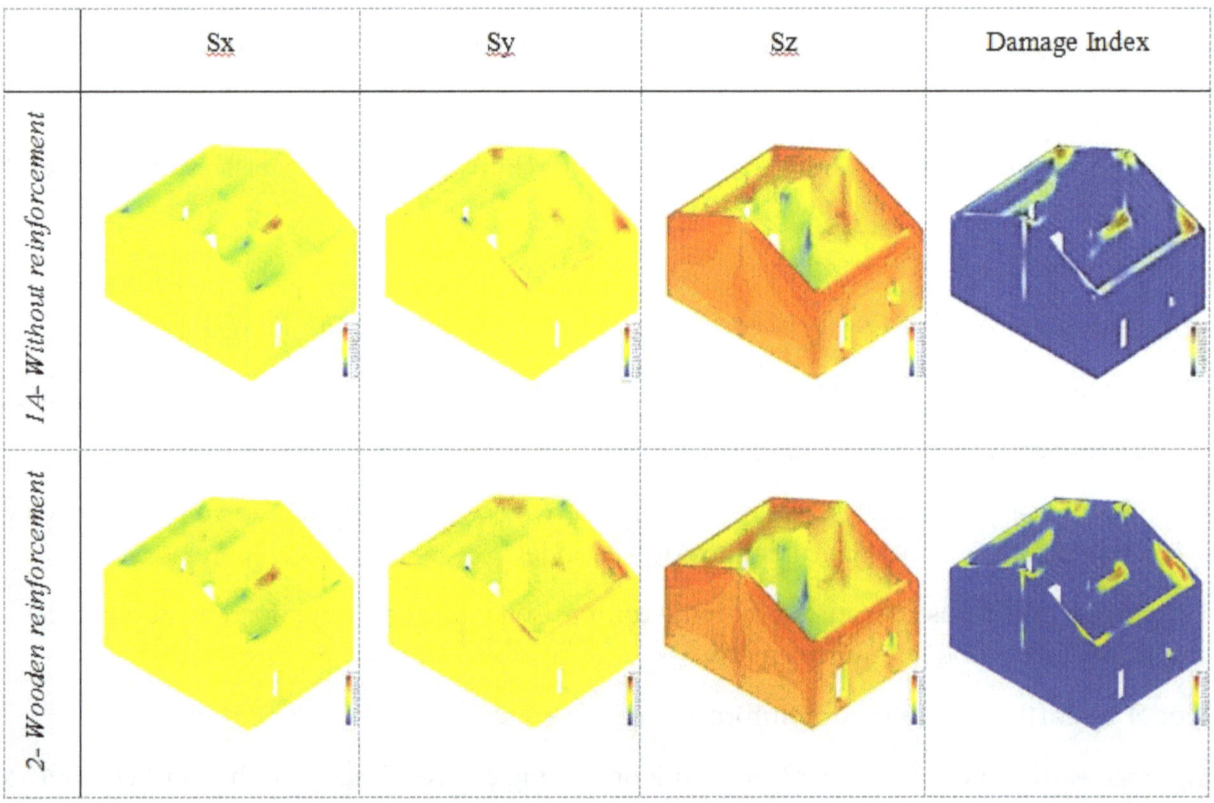

Figure 2. Static lineal method using gravitational and wind pressure.

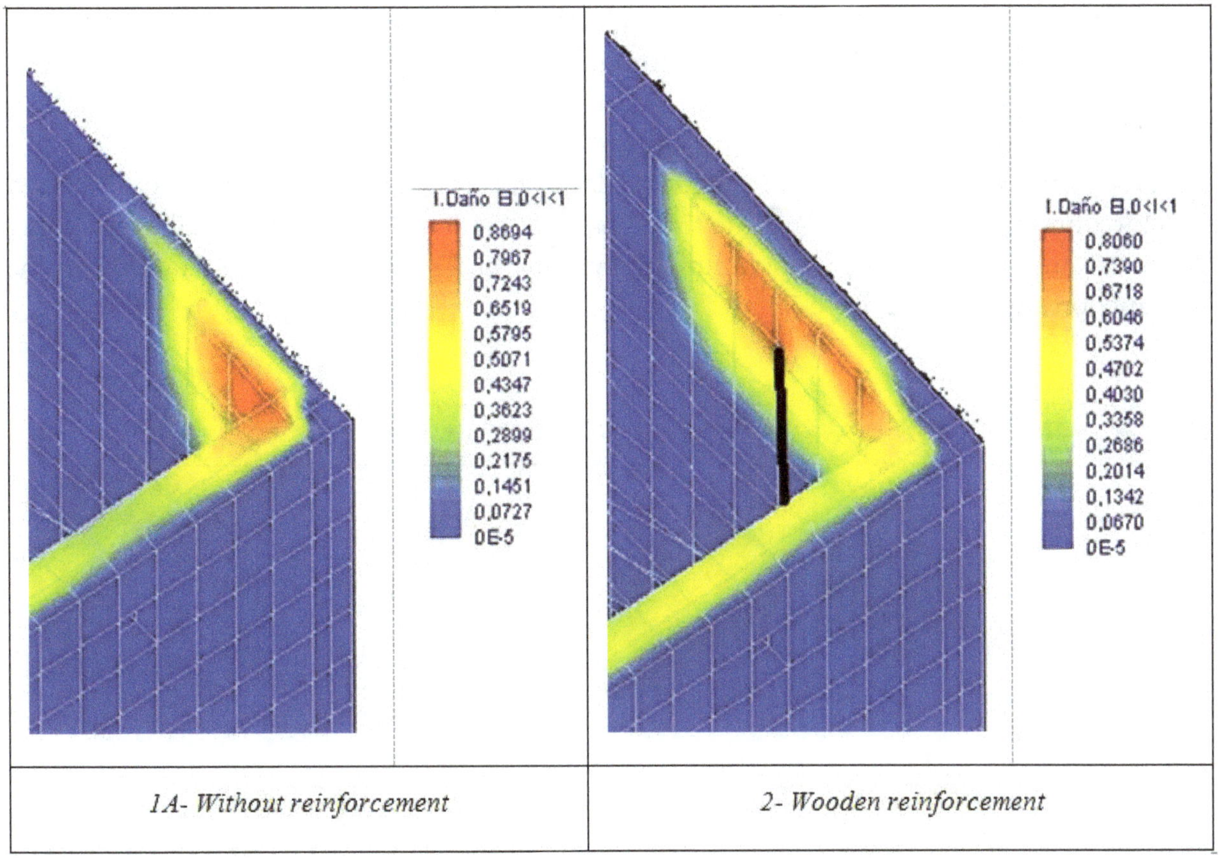

Figure 3. Static lineal method using gravitational and wind pressure. Enlarged view of wall joints.

produced where the reinforcement lies. The wooden reinforcement allows a redistribution of pressure and tension, thereby avoiding cracks in the superior wall joints.

- Non-linear static method, under the combination of gravity and horizontal loads until collapse.

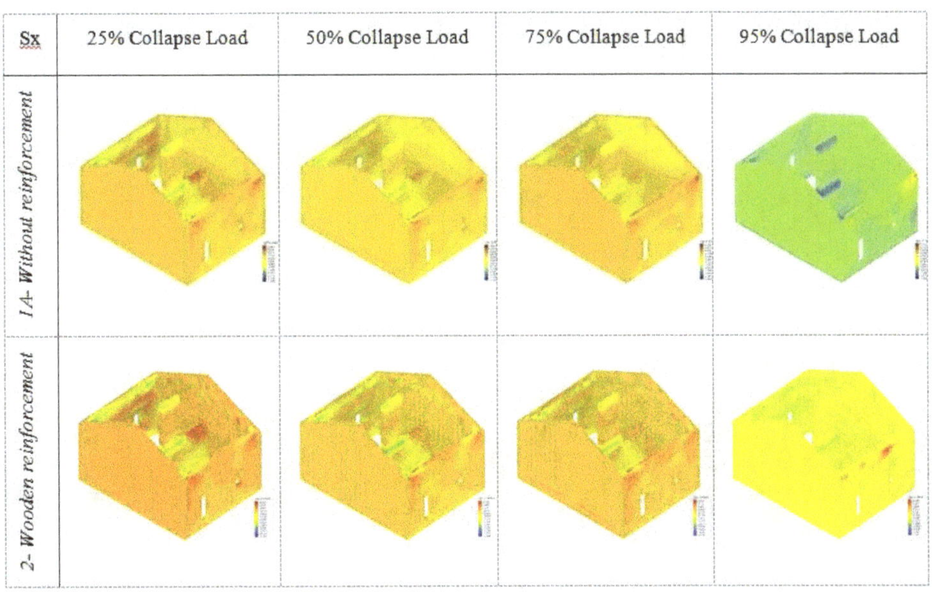

Figure 4. Non-linear static method, bearing gravitational and horizontal loads until collapse. Axis Sx.

In the graphs, with the consecutive increase of load (collapse load at 25%, 50%, 75% and 95%), there are consecutive increases of the pressure in the construction. (**Figures 4–6**).

- Dynamic-seismic method.

Under seismic conditions, evidently there are two types of common failures in earth-based structures and they would develop into the collapse of the building either by wall overturning failure or other unstable elements: failure by bending and shear failure (note failures in the introduction, Section 1) (**Figure 7**).

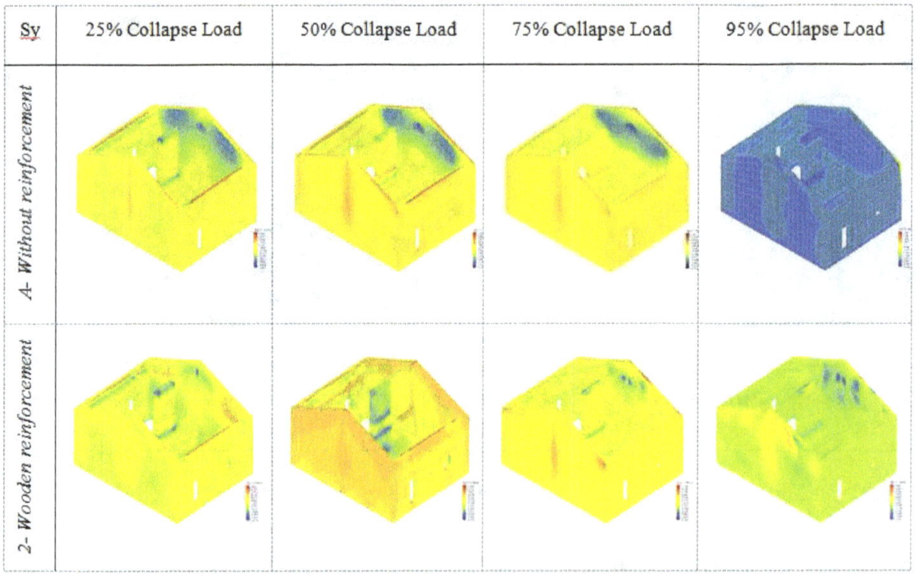

Figure 5. Non-linear static method, bearing gravitational and horizontal loads until collapse. Axis Sy.

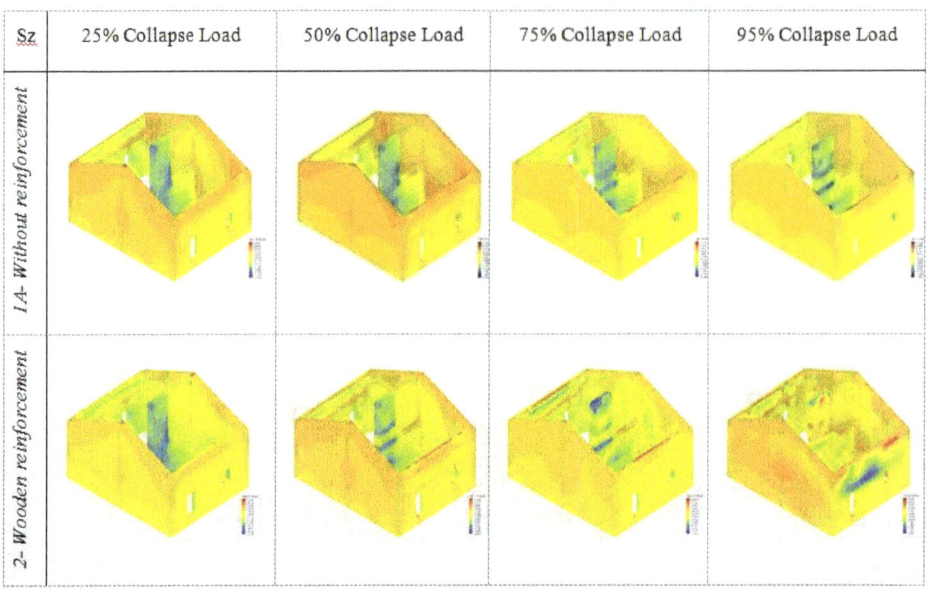

Figure 6. Non-linear static method, bearing gravitational and horizontal loads until collapse. Axis Sz.

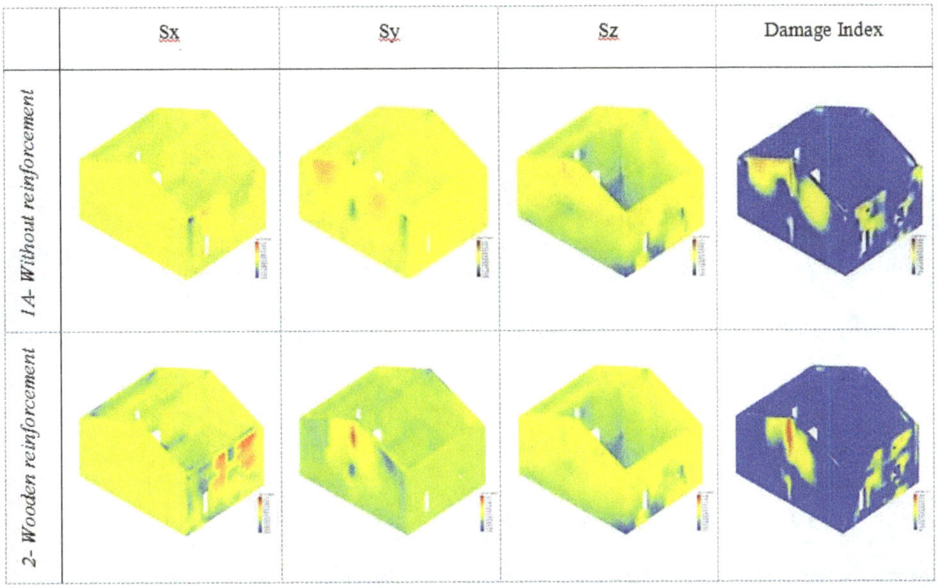

Figure 7. Dynamic-seismic method.

5. Damage rate

By calculating the pressures, we can acquire a damage index, which allows us to test the areas where the material no longer collaborates because it has been exposed to loads above its resistance capacity. This is especially interesting in the non-linear static calculation because of a combination of loads all gradually increasing; we can analyse all damage suffered as the load gets heavier. This way with the evolution of damage according to the increase in loads, we can study the response of the models as the loads increase according to the damage rate and see at what point the building will collapse.

Damaged material [10] normally comes from the top of the joints between the walls and progressively worsens as the load increases, thus causing the wall to collapse in two directions, thickness and height. The failure of the wall occurs when the fissures breach the wall completely and the walls become independent vertical cantilevers without lateral stability, so continuing to support loads will lead to collapse due to overturning failure.

From the wooden reinforcement model, we can see that the process is similar except that the reinforcements provide support between the walls increasing the collapse load capacity (**Figure 8**).

With same combination and increases, the exact load that collapses each model can be compared. 100% is the reference, maximum load buildings life, increasing loads until they fail, in so doing obtaining the collapse load for each model referenced in **Table 2**. With all the results and using model 1A as a reference, we can contrast the overall response of each of the models. This table gives a straightforward and direct comparison between the models analysed.

They show the different performances, assessing and quantifying their effectiveness and also graphics for a better understanding (**Table 2**) (**Figures 9** and **10**).

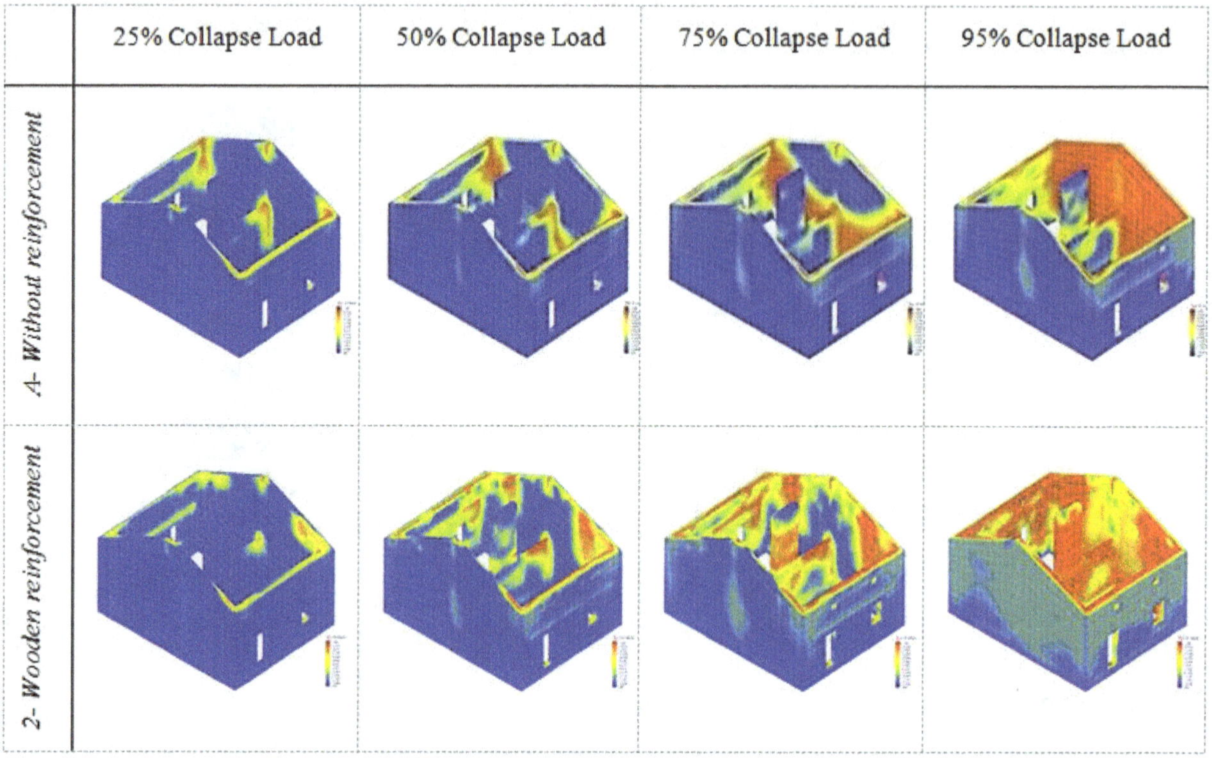

Figure 8. Evolution of damages according to increasing loads.

	Gravitational and horizontal (%)		Gravitational (%)	
	Collapse	Reference	Collapse	Reference
40-cm wall	280	100	600	100
60-cm wall	390	139	750	125
Wooden reinforcement	460	164	600	100

Table 2. Collapse load and coefficient breaking reference.

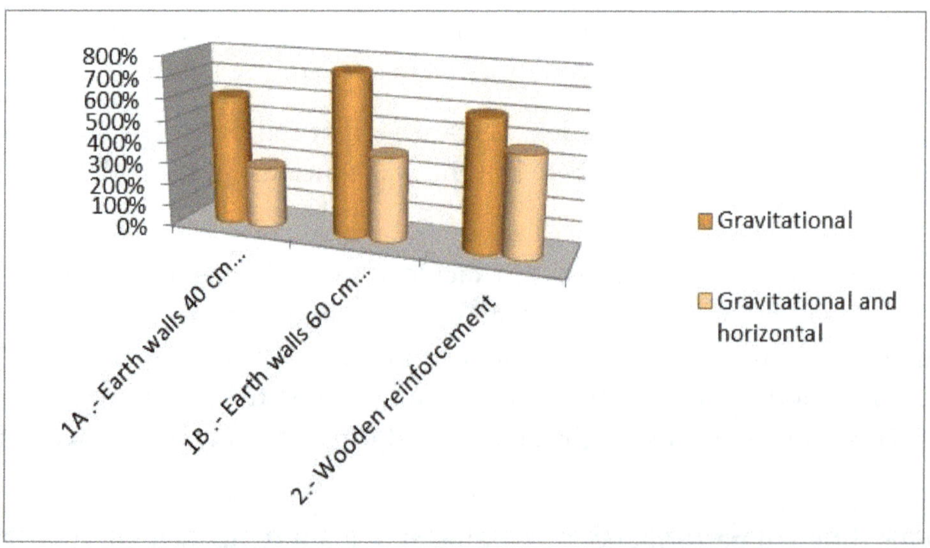

Figure 9. Collapse load.

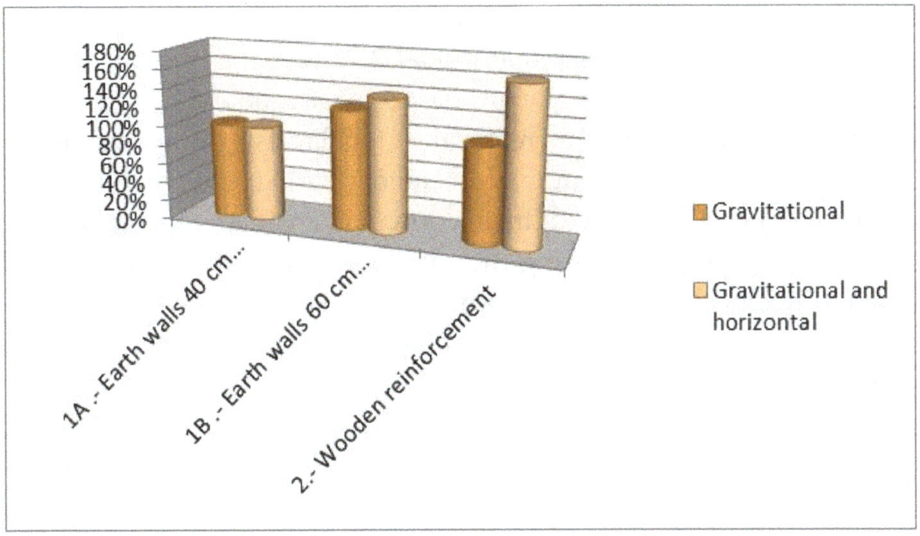

Figure 10. Coefficient breaking reference.

6. Results and discussion

In the models without reinforcements, the maximum tension is found in the uppermost corners of the walls. If these tensile forces are more than the resistance of the material, fissures will appear causing the walls to act as vertical cantilevers, and as a result beginning the process of building collapse due to a deficit in lateral wall stability. In model 1B, with a 50% thickness increase (40 cm to 60 cm), we see a considerable raise in global resistance of the structure. As predicted, an increased wall thickness dramatically increases its resistance in these areas and consequently its global structural strength [9]. Being the model with more area in its resistant section, in relation to gravity loads, this is the model with the highest resistance. This gives greater inertia against lateral overturning and also increases resistance to horizontal forces. Analysing the damage rate in the consecutive increase of loads, we can note that the fissure begins at the uppermost corner of the walls. From one side of the wall, the crack continues as particular wall sections buckle under the load, both transversely (wall width) and vertically (wall height). In the end, if the forces that created the initial split in the wall continue, the structure will fail. This process corresponds to the usual failure of the soil-based construction mentioned in Section 1, due to the low tensile strength of earth. Although with model 2, while wall separation occurs, the bracings give some reinforcement, permitting the walls to continue to work together, limiting the collapse due to failure in the joints. Owing to the effect of being tied, braced and joined to each other, this solution produces a significant increase in resistance against cracking at the uppermost part of the corners, resulting in a redistribution of tensile forces along the round pole timber, focused entirely at its joint. From **Table 2**. Collapse load and coefficient breaking reference, under gravity loads, the table shows that models 1A and model 2 has the same value, due to their walls being 40 cm thick. Model 1B highlight with an increase wall of 20 cm thickness (50% thickness) increases by 25% the total structural resistance against gravitational forces; the rational outcome of this being that vertical loads are transmitted through earthen walls. Thus, increasing their section will increase the resistant area and therefore its resistance to these forces. In the same way, under gravitational and horizontal pressure (wind), we can see that model 2—the global resistance of the structure increases substantially. Whereas the global resistance of model 1B increases 39%, placing wooden reinforcements into the walls increases the global resistance against these pressure by 64% more than the same model without reinforcements (1A). In spite of the round pole timber bracing solely interacting with the buildings' corners, it increases the collapse loads capacity significantly. This suggests that using the same material with which the slabs are built (thick round pole timber), contacting the upper corners to them, we can augment the global resistance of the structure significantly in comparison with the identical building without bracings, on the uppermost section where the walls join, areas in which earth structures are normally weakest. Developed models are studied and compared with existing structures. A complete series of vernacular houses still standing in the Castile area have been studied, but only a small number of them had walls with wooden reinforcements. Lastly, it is important to add that the cracks studied in the existing un-refurbished houses could be seen in the upper joints between the walls. This issue, in addition to the fact that these inhabited places have not been maintained, has caused a discoloration of the material surrounding the crack and therefore instability in not only the walls but also the roof, causing major problems and ultimately the house becomes ruins. The earth constructions with roofs and facades which have not been adequately looked after begin to gradually break down and decay. In the case of structures with wooden reinforcements, and adequate maintenance of the reinforcements, the roofs and the linings of the walls are free from cracks in the joints of the walls.

7. Conclusions

Soil-based structures, focusing on rammed earth walls and adobe, are susceptible to tensile loads, which are derived mainly from important horizontal external forces. This is amplified in the case of earthquakes: earthquakes with 0.20 g acceleration can bring earth-based structures without bracings to the verge of collapse; these kinds of earthquakes are common in regions where people continue to live and build earth constructions and where high seismic activity exists.

Common building failure occurs at the uppermost part of the joint between walls, causing them to become independent, losing lateral stability and giving way to collapse. Traditionally, reinforcing systems have been used with the aim of reducing this problem, which becomes more or less important depending on the building type and loading capacity. The solution of struts, round pole timbers in particular, increases the global resistance of the building significantly and also the collapse loads capacity against extreme horizontal forces. In terms of vertical pressures, increasing the wall section (50%) is the most adequate solution to increase the global pressure by 25%.

To increase the global resistance by 39% regarding horizontal pressures, one needs to increase wall thickness and employing round pole timber bracing, this benefit increases to 64% giving us the optimum solution not only structurally but also economically, requiring minimal materials and man-hours.

The detailing of the round pole timber bracings is left exposed (the pegs and wedges), allowing for a better control and maintenance during the building's lifecycle. Thus, it is recommended that the round pole timber bracing of the Castile region or a different reinforcement system be employed in all adobe structures to assure the stability of the walls. As indicated, the wall struts considerably increase their ability to withstand horizontal forces by creating an acceptable fixing between two walls. Bracing is a must for global stability and the monolithic nature of soil systems, and in seismic areas is a fundamental stabilising element.

Acknowledgements

This chapter presents some results of research conducted by the author in the journals "Constructions and Building Materials", "Journal of Architectural Conservation" and "Periodica Polytechnica Civil Engineering".

Author details

Q. Angulo Ibáñez

Address all correspondence to: quianib@csa.upv.es

Department of Architectural Constructions, Polytechnic University of Valencia, Valencia, spain

References

[1] Plinius Secundus, G. (Pliny the Elder), Naturalis Historia, 77.

[2] San Bartolomé, A., Curso de Albañilería Integral, PUCP, Lima, 2002.

[3] Minke G., Building With Earth. Design and Technology of a Sustainable Architecture, Birkhäuser Publisher Basel, 2013.

[4] Brzev, S., Tomazevic, M., Lutman, M., Bostenaru Dan, M., D' Ayala, D., Greene, M., The World Housing Encyclopedia: An Online Resource on Housing Construction in High Seismic Risk Areas of the World.

[5] Adell, J.M., Dávila, Mª.D., The Integral Masonry System. 13th IBMAC, Amsterdam, June 2004.

[6] Blondet, M., Vargas, J, Tarque, N., Velasquez, J., Experimental study of synthetic mesh reinforcement of historical adobe buildings, SAHC2006, Macmillan India Ltd. Vol 2, New Delhi, 2006.

[7] Font, F., Hidalgo, P., Arquitectures de tàpia, Collegi Oficial d'Aparelladors i Arquitectes Tècnics de Castellò, 2009.

[8] Graciani García, A, Barrios Padura, A, Tabales Rodríguez, M.A., Researches in Islamic Tapia Wall Construction in Southern Spain., Heritage, Weathering & Conservation. Londres, Reino Unido. Taylor & Francis Inc. 2006.

[9] Angulo Ibáñez, Q. Woodem reinforcement for earth constructions in Albacete, Spain. Journal of Architectural Conservation. England, 2015.

[10] Angulo Ibáñez, Q. et al, Solutions in traditional brace for rammed earth walls constructions in Manchuela albaceteña, Spain, 2012.

[11] Angulo Ibáñez, Q. et al, Traditional braces of earth constructions. Construction and Building Materials, England, Volumen 30, page 389, 2012.

[12] Angulo Ibáñez, Q. Performance analysis of wooden reinforcement in rammed earth walls. Periodica Polytechnica Civil Engineering, Hungary, 2017.

[13] Lonso Durá, A., Un modelo de integración del análisis estructural en entornos de Cad para estructuras de edificación, Tesis Doctoral, Departamento de Mecánica de los Medios Continuos y Teoría de Estructuras, Escuela Técnica Superior de Arquitectura, Universidad Politécnica de Valencia. Valencia, Diciembre, 2003.

[14] De Mazarredo Aznar, L., Calibrado de modelo de daño escalar con ensayos experimentales aplicado a materiales friccionales, Trabajo de investigación del Departamento de Mecánica de los Medios Continuos y Teoría de Estructuras, Escuela Técnica Superior de Arquitectura, Universidad Politécnica de Valencia. Valencia, Septiembre, 2011.

[15] Oller, S., Barbat, A.H., Oñate, E., Hanganu, A.D., A damage model for the seismic analysis of building structures, Earthquake Engineering, Tenth World Conference, Balkema, Rotterdam, 1992.

[16] Oñate, E., Structural analysis with the finite element method. Linear statics. Volume 1. Basis and solids, Springer, 2009. Volume 2. Beams, plates and shells, Springer, 2013

[17] Oñate E., Structural analysis with the finite element method. Linear statics. Volumen 2. Beans, plates and shells, Springer(2013).

[18] E. Oñate, J. Oliver, S. Oller, J. Lubliner, A constitutive model for cracking of concrete based on the incremental theory of plasticity. Engineering Computations (1988).

[19] Oller S., Oñate E, Oliver J, Lubliner J., Finite element non-linear analysis of concrete structures using a plastic-damage model. Engineering Fracture Mechanics. (1990).

[20] Oñate, E., Hanganu, A., Barbat, A., et al, Structural analysis and durability assessment of historical construction using a finite element damage model, CIMNE, Barcelona, 1996.

[21] Código Técnico de la Edificación, Documento Básico de Seguridad Estructural Acciones de la Edificación. España.

[22] NCSE-02, Norma de construcción sismorresistente: parte general y edificación. España.

3

Ductile Behavior of Timber Structures under Strong Dynamic Loads

Maria Cristina Porcu

Abstract

Due to their comparatively low mass that implies reduced horizontal dynamic loads even during strong earthquakes, wood-made buildings might be a good choice in seismic prone regions. To meet the modern design philosophy requirements, however, such structures should be able to behave in a ductile way under exceptional events. By presenting a brief review of the latest developments in the field, this chapter investigates on when and to what extent historical and modern timber buildings may exhibit a ductile and dissipative behavior. A special focus is given to the crucial role of connections and to the difficulties involved by their mechanical model when carrying out code-based non-linear dynamic analyses. Although a ductile behavior is typically required under strong earthquakes, it is to note that a well-designed ductile structure may also be able to withstand other exceptional events as, for instance, tornadoes or blasts.

Keywords: ductile behavior, strong dynamic loads, timber structures, seismic performance

1. Introduction

To be able to exhibit a ductile behavior is a bonus that a structure can exploit when exposed to extreme and unpredictable actions like those applied by earthquakes, hurricanes, snowstorms, fire or explosions. In fact, it would be uneconomical to design structures that withstand such infrequent events still deforming in the elastic range. In addition to cost-effective reasons, a significant advantage of the ductile behavior is the great amounts of input energy that can be dissipated during the plastic motion, considerably higher than what could be dissipated in the elastic range. Further advantages are the beneficial redistribution of stress within statically undetermined systems and the warning that large deformations may give to inhabitants in case of exceptional loads. Aimed by these reasons, the seismic codes of practice typically require that buildings survive earthquakes of moderate severity deforming in the elastic range whilst they behave in a dissipative ductile way, even getting damaged severely but without collapsing, during strong seismic events. Local curvature ductility and strength hierarchy requirements ensure such extreme performance of buildings with the final goal of life safety.

Based on this well-established design philosophy, a lot of experimental data, numerical procedures and code provisions are in fact available to design earthquake-resistant buildings made by steel or reinforced concrete [1–3]. On the contrary, a lack of sufficient research results and detailed code requirements still affect the design of wood structures in high seismicity areas, despite the growing interest in natural and renewable materials. Light weight as they typically are timber structures might possibly resist earthquakes better than other kind of structures. Being proportional to the building masses, in fact, the horizontal loads acting on light structures during a given earthquake are lower than those acting on heavier structures, so that reducing the building active mass may be even a strategy for seismic stress control [4, 5].

On the other hand, to meet the modern design philosophy, timber structures are expected to exhibit a ductile and dissipative behavior—besides adequate strength and stiffness—under strong earthquakes. Ductility is in fact the capability to undergo large inelastic displacements without significant reduction in strength. On the other hand, high values of ductility usually imply large amount of energy dissipated during the dynamic motion, being the energy dissipated in each cycle related to the hysteresis area of the load-displacement diagram.

With the aim of investigating on the dynamic behavior of timber structures under extreme conditions, this survey focuses on the post-elastic behavior of wood constructions. The chapter is composed of two main parts. The first one deals with the ductile behavior of historical buildings (Section 2). A brief review on the performance of existing wood-made buildings under past events is presented in Section 2.1 while experimental data on the ductile and dissipative behavior of ancient timber buildings, as available from the literature, are provided in Section 2.2. The remarks listed in Section 2.3 close the first part of the chapter. The second part refers to the ductile and dissipative behavior of modern timber constructions (Section 3). The key role of metallic connections is addressed in Section 3.1, whereas some practical aspects to be taken into account before carrying out a code-based non-linear dynamic analysis of a timber structure are highlighted in Section 3.2. Some brief conclusive notes are provided in section 3.3.

2. Ductile and dissipative behavior of ancient timber buildings

Until the advent of reinforced concrete, wood buildings have been popular for centuries in many seismic areas like China, Japan, Greece, Turkey and Balkan countries. In other geographical areas prone to earthquakes or tornados (North America), wood-frame single-family houses and low-rise multi-family dwellings never stopped to be built since colonial times. Timber constructions are today still very common in countries where wood is in good supply (USA, Canada, North Europe and New Zealand), but a growing interest is developing even in regions where adobe and masonry have been the most typical construction materials for centuries (e.g. Mediterranean countries). Assessing the performance of timber structures during past severe natural events may be thus of great interest not only to preserve the architectural heritage but also to improve the design of new buildings.

2.1. Performance of timber buildings during past natural events

Somehow unexpected, the finding that many ancient wooden constructions remained nearly intact after strong earthquakes, whilst modern reinforced concrete buildings collapsed, as emblematically shown in **Figure 1**, gave new impetus to build timber constructions even in seismic regions. Several authors documented, in fact, a generally good performance of timber buildings during natural events as earthquakes, hurricanes, tornadoes and snowstorms [6–12].

Actually, most of the existing timber constructions were built according to conventional guidelines based on the empirical knowledge of the past while only the more recent ones (a little part) are engineered structures meeting modern anti-seismic standards. Mainly based on inspections made after natural disasters, best-practice rules were suggested—although not always applied—in the past [13]. It may be interesting to mention that the first European anti-seismic recommendations, which were enacted in 1783, after a devastating earthquake struck Calabria and Sicily (at that time belonging to the Borbone Kingdom), prescribed to strengthen buildings by means of wood frames in filled in the masonry walls. The timber frames were recommended to be suitably connected to each other's, to the ground and to the floors, see **Figure 2** [14]. Even more interesting is the fact that such an aseismic system (adopted in Italy till the twentieth century and referred to as "*casa baraccata*"), had actually been borrowed from the traditional little wooden single-story houses (called *baracche*) which were usually built near to the palaces of nobles to be used as earthquake-proof shelters [15–17]. Although not written, aseismic recommendations similar to—and most likely inspiring—the Borbone's ones were adopted in Lisbon a few years before, after the catastrophic earthquake and the ensuing tsunami that struck the town in 1755. Based on the knowledge collected from survived buildings, the *Pombalino* constructive system (imposed by the Prime Minister Marquis of Pombal) consisted in fact of a three-dimensional timber structure (called *gaiola* due to its cage-like structure) included in the masonry walls above the first floor of the building, see **Figure 3** [18–20].

Figure 1. Different damaged conditions of traditional wooden and reinforced concrete buildings after the 1999 Duzce earthquake. Image taken from [6].

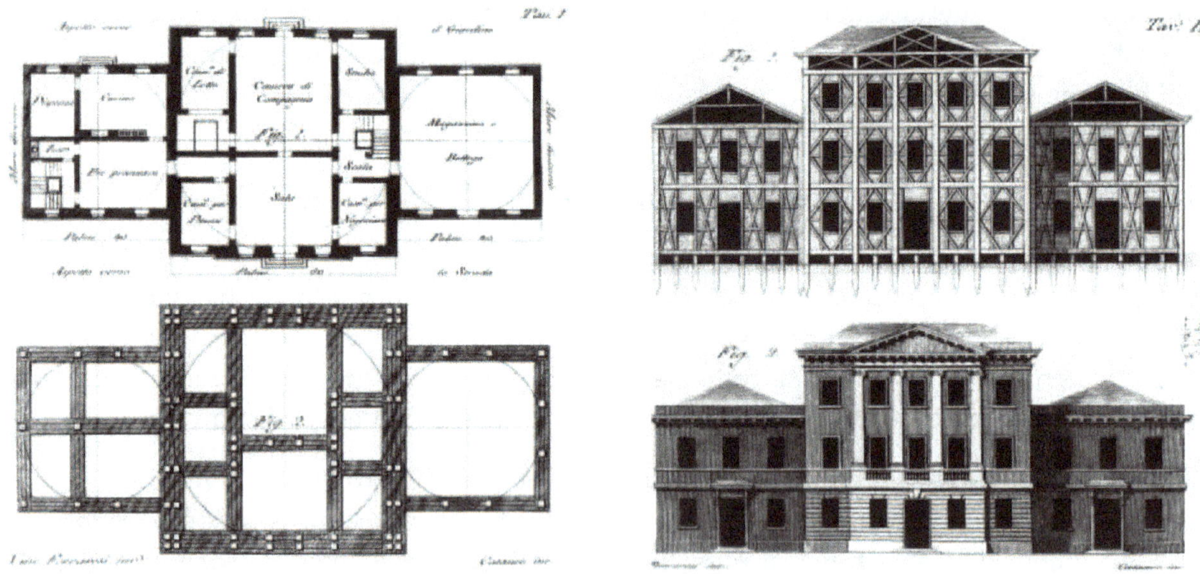

Figure 2. First European aseismic recommendations dating back to 1783 [14].

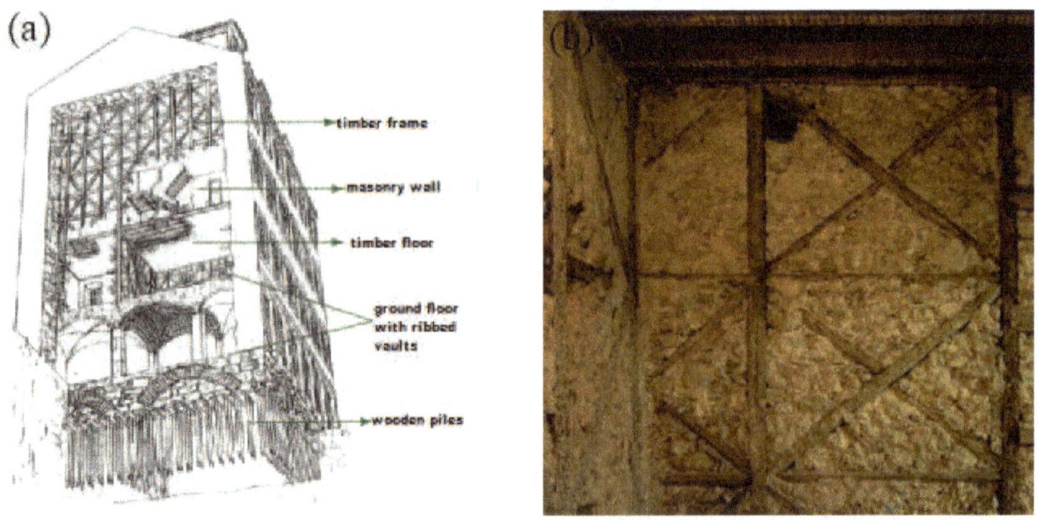

Figure 3. (a) Pombalino building scheme [18]; (b) Pombalino building walls.

Due to the different role given to the timber structure, the "Borbone" and the "Pombalino" constructive systems were quite dissimilar in their mechanical behavior, as discussed in Ref. [16]. Both of them are, however, examples of ancient timber-framed buildings. Although the concept of structural ductility was still very far to be introduced in building design, the role of timber frames in traditional constructions was in fact that of withstanding lateral loads by allowing large—and usually inelastic—displacements (like a sort of reinforcement skeleton), a role which cannot be played by unreinforced masonry.

Constructive systems adopting timber frames with infill masonry (or adobe or stone) may be actually found in different historical periods and geographical areas, starting from one of the most ancient examples still visible: a Roman house of the first century BC in the archeological site of Herculaneum (Italy) (**Figure 4a**). More recent examples are the several well-preserved

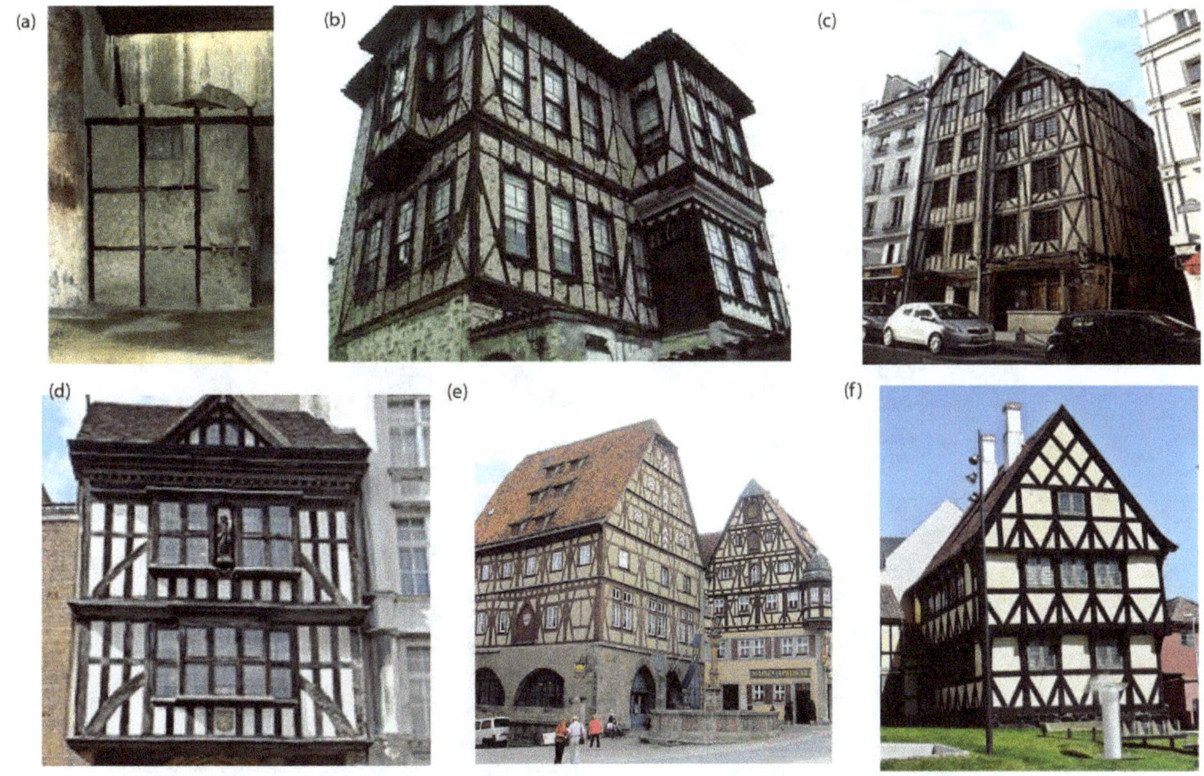

Figure 4. Historical timber-framed masonry buildings: (a) house wall from Herculanum (opus craticium attested by Vitruvius in the first century BC); (b) nineteenth-century Ottoman house; (c) sixteenth-century timber-frame house in Paris (France); (d) St. Bartholomew's Gatehouse built in London in 1595 and survived to the 1665 great fire; (e) German typical timber-framed houses (Fachwerkhäuser); (f) Anne Hvide's House built in 1560, Denmark.

Ottoman traditional timber-frame houses (see **Figure 4b**), that may be still found in earthquake-prone areas as Turkey, Balkan countries, Iraq, Siria and Egipt [6, 9, 21]. Likewise the *Pombalino* system, also in the Ottoman houses, the timber frames were introduced from the first floor up (for this reason they are also referred to as half-timbered frames). Further examples of timber-framed buildings (dating back even to the medieval period) may be also found in European countries as Germany, the United Kingdom, Greece, France, Spain, Poland (see e.g. the examples given in **Figure 4c–f**), as well as in India [22, 23], the USA [24], China [25] and Peru [26]. An overview of the history of timber framed buildings may be found for instance in Ref. [32].

Different from the timber-framed systems, other wooden ancient buildings can be also broadly found (see **Figure 5**).

The widespread use of wood as structural material even in earthquake-prone areas highlights the awareness they had in the past of the good performance of such a material even under extreme conditions. Of course, mistakes and experience taught how to build timber structures to better withstand strong dynamic actions. The following shortcomings were typically found, in fact, to be responsible for the inefficient structural behavior of wood buildings in past natural disasters: lack of adequate foundation anchorages or inadequate foundations soil; presence of a soft-story mechanism (typically due to large openings at the ground floor of multi-story buildings); insufficient connections between major components or lacking lateral bracing, as

Figure 5. Ancient wood constructions: (a) Hōryū-ji Pagoda (Japan, 607 A.C.); (b) Kizhi Pogost Church (Russia, seventeenth century); (c) Fogong Temple Pagoda (China, 1056 AC); (d) Heddal stave church (Norvey, thirteenth century).

documented, among others, in Refs. [6–11]. Other causes of failure were also detected as, for instance, failure of infill materials in timber-framed buildings, failure of chimneys, collapse of neighbor buildings, structural changes or excessive loadings given at a certain stage of the building life which caused very large inertia forces [6, 9, 11, 27, 28]. However, the main lesson drown from past events was that, to better resist to lateral loads, a timber structure should be regular, have little openings, possess effective connections, bracing elements and ground anchorages. In other words, it should eventually behave as a box-like unit.

2.2. Ductility of historical buildings

Apart from the qualitative considerations recalled in the previous section, we can ask whether and to what extent existing (and often very ancient) timber structures can be able to exhibit a ductile and dissipative behavior under extreme conditions. Unfortunately, answering this question is not an easy task, since the actual dynamic behavior of historical timber buildings is hardly predictable and it is usually documented, as far as for the past, only by empirical data. The task is even made more difficult by the different constructive techniques, wood species, kind and quality of infill materials and by the large variety of joints adopted in ancient structures. Some recent studies can be, however, exploited to have some clues.

2.2.1. Timber-framed buildings

Experimental researches were carried out to assess the behavior of the *Borbone* constructive system, [16, 17, 29]. Particularly remarkable are the results provided in Ref. [16], relevant to in-plane cycling tests performed on two full-scale models of *Borbone* walls, following the UNI EN 12512:2006 protocol [30]. Experimental findings evidenced a non-linear behavior of specimens with comparatively high values of ductility. The latter was calculated as the ratio between the maximum displacement $u_{\max}$ and the displacement at yield u_y, namely

$$\mu = \frac{u_{\max}}{u_y} \tag{1}$$

The quantity defined by Eq. (1) is sometimes also referred to as the static ductility. To quantify the dissipation of energy in the plastic range, the hysteresis equivalent damping ratio was calculated in Ref. [16], as defined by

$$\xi_{eq} = \frac{1}{2\pi}\frac{E_d}{E_p} \tag{2}$$

Based on the Jacobsen approach [31], Eq. (2) considers the ratio between the energy dissipated for a half hysteresis cycle, namely E_d, and the potential energy E_p that an equivalent simple oscillator would store for the same displacement achieved in the hysteretic half-cycle (see **Figure 6**). The value of ξ_{eq} was evaluated in Ref. [16] for the third cycle of each level of displacement imposed, as suggested by the UNI EN 12512:2006 protocol [30]. The cumulative dissipated energy E_{d_cum}, obtained by summing the loop areas of successive cycles, was also calculated. The values of μ, E_{d_cum} and ξ_{eq} obtained in Ref. [16] for infilled and bare walls are provided in **Table 1**.

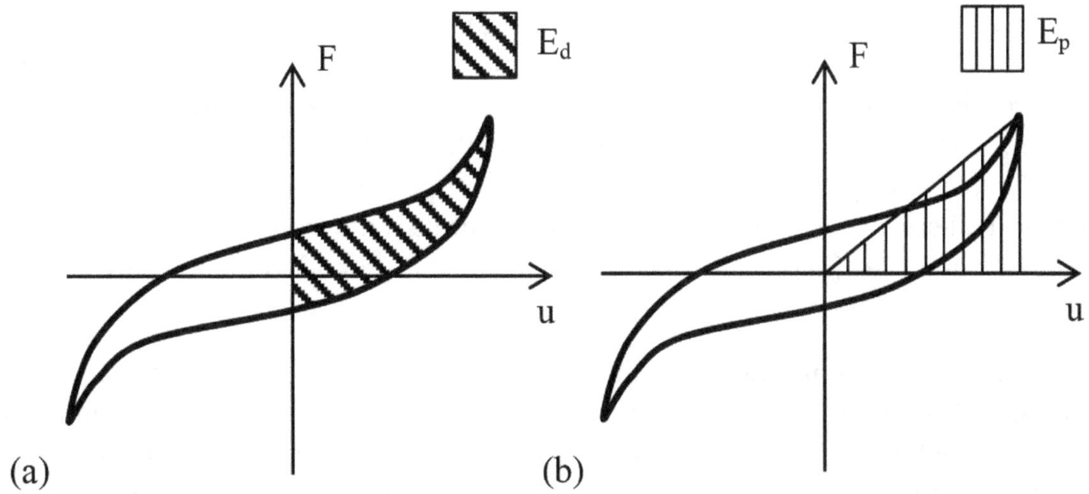

Figure 6. Pinched hysteresis loop (third cycle): (a) dissipated energy for a half cycle; (b) potential energy of the equivalent oscillator for a half cycle.

Test specimens	Frame	μ	E_{d_cum} (kJ)	ξ_{eq} (%)	Drift (%)
Real-size *Borbone* walls [17]	Infilled	7.6	19	6–8.9	2–2.6
	Bare	1.36	1.6	1–23	0.15–2.71
Real-size *Pombalino* walls [32]	Infilled	5.48–17.95	2–2.8	15	1.76–3.34
	Bare	11.6–15.28	0.75–2.8	11–15	2.55–2.61
Real-size Ottoman walls [21]	Infilled	n.a.	1.8–11	7.80	4–6.5
	Bare	n.a.	0.6–8	7.79	3.5–8.7
Real-size *dhajji-dewari* walls [23]	Infilled	3.7	n.a.	5–25	4.2–5
	Bare	3	n.a.	7	4.2–5
Half-scale *quincha* walls [26]	Infilled	6.2	n.a.	n.a.	n.a.
	Bare	15.4	n.a.	n.a.	n.a.

Note: n.a., not available.

Table 1. Ductility ratio and dissipation capacity of historical timber-framed walls.

In the infilled specimens, the energy dissipation was found to depend mostly on friction and small ruptures in the masonry. In bare specimens, the almost wholeness dissipation was found instead to be concentrated in the connections, where two mechanisms occur contemporarily: the first one is related to the compression and crushing of wood grain with the formation of a cavity due to the presence of nails (which is also responsible for the pinching effects discussed in the following); the second one concerns the inelastic deformation of the iron nails, [16, 17].

The dynamic behavior of the *Pombalino* timber-framed buildings in Lisbon was investigated through experimental tests on specimens taken from actual sites [28] or rebuilt in laboratory [19, 27, 32]. All of them assessed a rather good ductile behavior of *Pombalino* walls. In particular, the results given in Ref. [19], relevant to real-size reconstructions of *Pombalino* walls under cyclic displacement sequences following the CUREE protocol [33], showed fat hysteresis loops dissipating reasonable amounts of energy (although the actual values are not given by the authors). The ductility ratio was found to be around 3. Significant pinching effects, related to gaps and residual displacements which form in the connection zones and between masonry and timber members were detected. A rather accurate non-linear hysteresis model was also developed in the same paper to capture the dynamic response of the walls. Similar experimental tests on real-size laboratory reconstructions of infilled and bare-framed *Pombalino* walls were carried out in Ref. [32]. The values of μ, E_{d_cum} and ξ_{eq} found in Ref. [32] are collected in **Table 1**.

Experimental tests on eight different timber frames with and without infill (brick and adobe) or cladding, all reproducing portions of the typical *himis*-timber-framed Ottoman constructions, were carried out in Ref. [21]. The force-displacement curves obtained for reversed cyclic in-plane lateral loading showed stable hysteretic loops with highly dissipative behavior. The authors suggested to evaluate the cumulative energy dissipation capacity through the following empirical formulas [21]:

$$E_{d\_cumW} = (0.873W_{eff} + 0.055A_{eff})(1.1865D_{\max} + 0.5697)^{3/2} \tag{3}$$

$$E_{d\_cumWO} = (W_{eff} + 0.0638A_{eff})(0.19D_{\max} + 1.0469)^{2} \tag{4}$$

Here E_{d_cumW} and E_{d_cumWO} denote the cumulative energy dissipation capacity (expressed in Joule) of specimens with and without infill/cladding, respectively; while W_{eff}, A_{eff} and $D_{\max}$ are the effective width (m), the effective area (m^2) and the maximum lateral top displacement (mm) of each framed wall considered in the experimental tests. The values of E_{d_cum} as obtained by applying Eqs. (3) and (4) and of ξ_{eq}, as defined by Eq. (2), were derived from the data given in Ref. [21] and are provided in **Table 1** both for infilled and bare frames. A high ductile behavior was found to be exhibited by the considered timber-framed walls, although the actual values of the ductility ratio μ were not provided in Ref. [21]. The highly dissipative behavior detected for Ottoman walls was found to depend on the dissipative plastic behavior of nailed connections, as also confirmed by other authors [22].

The results of in-plane quasi-static cyclic tests on full-scale *dhajji-dewari* walls (wooden-braced frame system with masonry infill, typical of India and Pakistan) were given in Ref. [23]. The tests highlighted a very strong resilience of the *dhajji-dewari* system against lateral loads, which is almost totally due to the performance of the timber framework. Also in this case, a rather good ductile and dissipative behavior is found in the experimental tests. The values of μ and ξ_{eq} obtained in Ref. [23] for infilled and bare walls are given in **Table 1**.

Finally, the data taken from [26] relevant to experimental tests carried out on half-scale *quincha* walls are also provided in **Table 1**. (*Quincha* is a Peruvian traditional timber-framed constructive system consisting of an adobe-made ground floor and upper stories built with timber frames infilled with a weave of canes and mud).

A qualitative diagram is provided in **Figure 7** showing some of the typical features evidenced by the experimental cyclic tests on timber-framed walls, infilled with masonry and connected with metallic nails. The main features evidenced in Figure 7 are (a) non-linear behavior; (b) almost symmetric curves; (c) indistinct yield point; (d) lateral stiffness degradation for increasing loading cycles; (e) pinching effect after the first load cycle; (f) strength degradation at the same deformation level for increasing loading cycles; (g) strength degradation for higher deformation;(h) fat hysteresis loops which implies large amounts of energy dissipated (narrowed loop areas are generally found, however, for successive load cycles or for higher deformation levels); (i) rather high values of ductility.

The so-called "pinching effect" is due to the formation of a cavity around the fasteners due to irrecoverable crushing of wood (this effect occurs after the first loading cycle) which implies a reduced stiffness at load reversals (the connection stiffness is due, in this phase, to the sole contribution of metallic fasteners). As soon as the contact with the surrounding wood is reestablished, at increased deformation levels, the stiffness rapidly increases which leads to the typical pinched shape of curves in the load-displacement diagram. It can be noted, finally, that features very similar to those illustrated in **Figure 7** and discussed before were

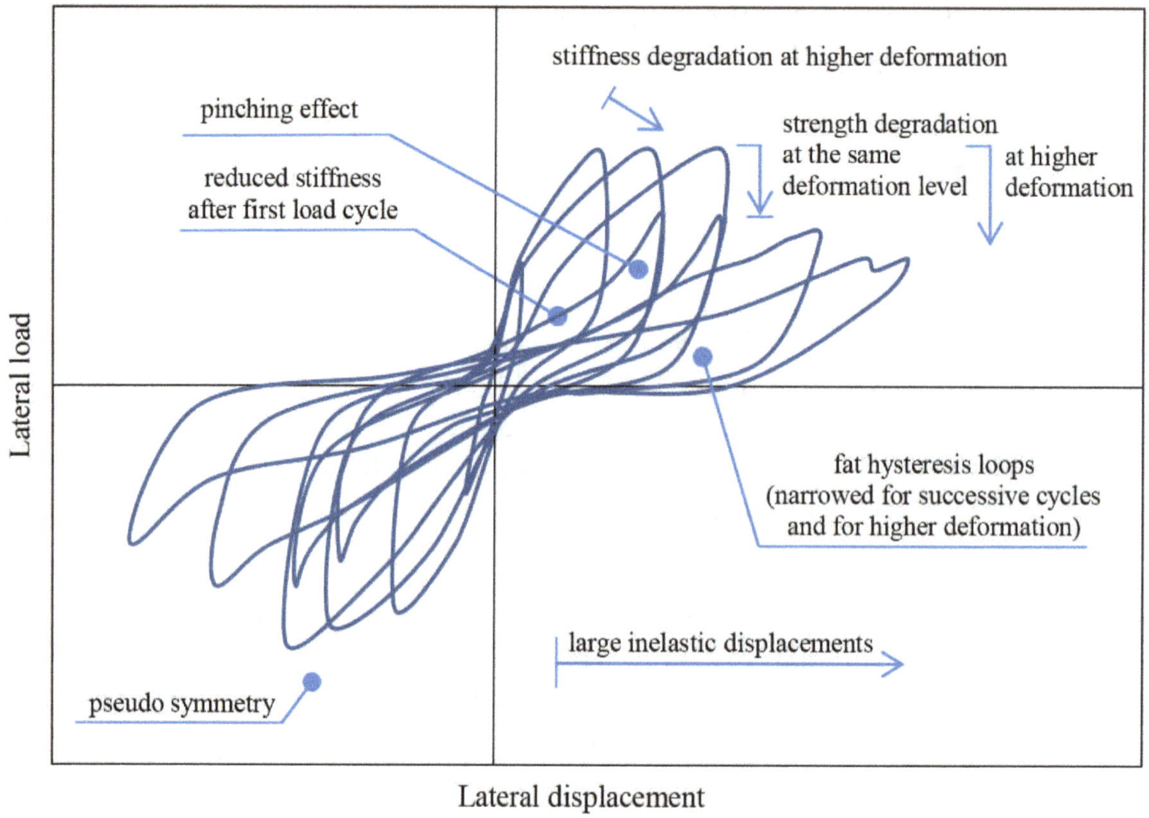

Figure 7. Qualitative hysteresis loops as obtained from experimental tests on timber framed walls.

also detected for modern timber walls (plywood shear, strand-board diaphragms), as can be inferred by examining, for instance, the experimental load-displacement curves provided in Ref. [34].

2.2.2. Other kinds of ancient wooden constructions

Experimental studies were carried out to assess the dynamic behavior of traditional Chinese [35–37] and Japanese [38–40] constructions. Of particular interest to the purpose of the present study are the results provided in Ref. [37]. They are relevant to scaled specimens, reproducing a prototype of a historical Chinese timber palace, subjected to cyclic lateral loading. Stable and large-area hysteretic loops were found in the tests, associated with rather high values of ductility (9–19) and energy dissipation (not quantified). The equivalent viscous damping was evaluated to reach even values of 20–30%. It is worth noting that Chinese and Japanese constructions were found able to exhibit a rather high ductile response and to dissipate large amount of energy, despite they typically avoid metallic connections. Energy dissipation was found to be due, in fact, to friction and embedding between structural elements at the contact interfaces inside the wooden joints.

Further experimental studies can be also quoted, for example [41], confirming a rather good ductile and dissipative behavior of ancient timber constructions.

2.3. Some remarks on the ductility of ancient timber constructions

Before closing the present section, the following remarks can be drawn.

i. Ductile behavior and energy dissipation are two key points of the modern aseismic strategy adopted by current standards. Experimental findings showed that historical timber buildings can generally be able to meet these modern requirements.

ii. **Table 1** substantiates statement (i). With reference to the classification given by EC8 [1], recalled in **Table 2**, three kinds of the traditional timber structures referred to in **Table 1** may be assigned to the high ductility class (DCH), namely the *Borbone*, the *Pombalino* and the *quincha* walls (all in the case of infilled timber frames). A low ductility class (DCL) should be assigned instead to the *dhajji-dewari* walls. The class of ductility of the Ottoman traditional walls cannot be assessed through the available data.

iii. Although giving an interesting portrait of the ductile and dissipative capacity of traditional timber buildings, the data collected in **Table 1** should be compared with caution due to the different settings (laboratory set-ups, loading procedures, recorded data analysis, yield and ultimate deformation evaluation) adopted in the experimental studies mentioned. For instance, the total amount of energy dissipated can be strongly dependent on the load protocol, which is usually different from an experimental research to the other. In addition, it can be noted that a different ductile behavior was sometimes detected in experimental tests from positive to negative direction of load, although such an aspect has not been evidenced in **Table 1** for the sake of brevity.

iv. The last column of **Table 1** reports the values of the drift (in %) as given by the ratio between the maximum lateral displacement and the wall height. This is in fact a very significant parameter to be considered when large displacements are involved.

v. Despite the extreme attention paid by the authors in reproducing, as faithfully as possible, the in-situ conditions(timber species, infill materials, geometry, ground constraints, connections, and so on), the results obtained in laboratory on rebuilt models of parts of ancient constructions should be obviously used with great care to predict the actual dynamic behavior of existing whole buildings, also in view of all the aspects affecting the behavior of real structures (material degradation, efficiency of connections, internal damage, workmanship irregularities, tridimensional behavior of the building).

vi. A crucial role in the ductile behavior of ancient timber structures was found to be played by connections, since timber elements generally do not exploit their latest strength resources [21, 42]. Based on this statement, some retrofitting solutions were also proposed to improve the dynamic behavior of ancient structures under dynamic loads [27, 42]

3. Ductile and dissipative behavior of modern timber structures

Besides the more common single-family and low-rise houses, spectacular and daring-shaped modern timber buildings may be even encountered nowadays in many countries, as the few

μ	Class of ductility	q	Structural type
≤4	DCL Low capacity to dissipate energy	1.5	Cantilevers; beams; arches with pinned joints; trusses joined with connectors; mixed structures consisting of timber framing and non-load bearing infill.
4<μ<6	DCM Medium capacity to dissipate energy	2	Glued wall panels with glued diaphragms connected with nails and bolts; trusses with doweled and bolted joints.
		2.5	Hyperstatic portal frames with doweled and bolted joints possessing medium capacity of rotational ductility.
μ≥6	DCH High capacity to dissipate energy	3	Nailed wall panels with glued diaphragms, connected with nails and bolts; trusses with nailed joints
		4	Hyperstatic portal frames with doweled and bolted joints possessing high capacity of rotational ductility.
		5	Nailed wall panels with nailed diaphragms, connected with nails and bolts.

Note: The q-values should be reduced by 20% if the building is non-regular in elevation.

Table 2. Ductility classes and behavior factors q for different typologies of timber structures, according to EC8 [1].

instances of **Figure 8** let imagine. A feeling for eco-friendly and renewable materials, together with the easy of production and transportation from the past, adds new motivations to the construction of wooden buildings.

As discussed in the introductive section of this chapter, modern structures are required to be ductile and dissipative, particularly when they are built in seismic areas. While timber structures are uniquely recognized to be able to meet such requirements provided that they are

Figure 8. (a) Dolomites, Italy; (b) Vancouver, Canada; (c) London, Great Britain; (d) Odate Stadium, Japan.

regular, hyperstatic and connected with ductile fasteners (as also confirmed by **Table 2**), most of the issues related to evaluating and modeling this ability are still under discussion.

3.1. Crucial role of connections

Connections in modern timber buildings are metallic devices ensuring transmission of forces between structural elements. Their design is the most strategic part of the structural project of a timber construction, since from the characteristics of the connections (type, mechanical properties, geometry, spacing, assembly techniques) may strongly depend the stiffness, the strength, the ductility and the energy dissipation of the whole structure.

Although some constructive typologies (such as moment-resisting timber frame systems, timber shear panel systems and cross-laminated panel systems) are indicated as being particularly able to ensure a ductile behavior under extreme dynamic lateral loads [43], it is the connection design that eventually decides the ductility resources of a timber structure. The same structural type may be, in fact, assigned to different ductility classes in dependence of the rotational ductility capacity of its connections, as can be inferred, for instance, by the classification done by EC8, recalled in **Table 2**.

The most common connections in modern timber structures are the dowel-type mechanical fasteners (nails, screws, dowels, bolts, rivets) which deeply penetrate into the wood to transfer the load by means of wood bearing and connector bending. Dowel-type connectors can be used alone or in combination with metal predrilled plates. Joints with dowel-type fasteners are expected to be ductile due to the highly nonlinear behavior of the wood under embedding stresses and the plastic behavior of the steel fasteners in bending [44]. Nevertheless, they can sometimes be affected by sudden and brittle failures like block shear or splitting [45]. Ten different types of failures (six in single shear and four in double shear) are considered by the European standards for dowel-type timber connections [46].

As a matter of fact, timber members and metallic joints play different roles in the seismic behavior of timber structures. Since the failure mechanisms of wooden elements are mostly brittle, the timber members are required to remain in the elastic range even under very strong events. The task of satisfying the demand of ductility is entrusted instead to the metallic connections which are expected to sustain large inelastic deformations while preventing collapse. The ductile behavior of connections is influenced both by metallic fasteners (which may behave in a ductile or brittle way depending on whether plasticization is attained or not) and by the strength properties of the wood surrounding the connection zone (direction of the grain with respect to the load direction).

Preventing brittle failure may guarantee an adequate ductility to the whole structure. Complying some strength hierarchy rules can assure a ductile behavior to timber structures. In particular, it is essential to design the fasteners to be weaker than the wood members they are connecting, so that they can yield and dissipate great amount of energy. On the other hand, the weaker the fasteners, the lower their bearing capacity. A way of ensuring both adequate ductility and sufficient bearing area is using a large number of weak fasteners. Some alternatives to improve the performance of dowel-type joints are discussed in Ref. [47].

Although the plastic properties of the steel fasteners alone are well-known and their behavior under cyclic loads easy predictable, the non-linear response of the assembly of metallic connectors and surrounding wood is rather difficult to predict, since it is not a cross-section property (as for reinforced concrete). In fact, the behavior of the timber connections depends from several factors, some well-known as the strength properties and the geometric configuration of involved materials, others affected by uncertainty as the influence of neighboring metallic fasteners or the interaction between fasteners and surrounding wood. This makes rather difficult to develop an analytical model able to reproduce the behavior of a timber connection.

Most of the features evidenced in **Figure 7** and discussed in Section 2.2.1 characterize the behavior of metallic timber connections, as can be inferred from **Figures 9a** and **9b**, which provide qualitative examples of the typical hysteretic behavior of riveted and nailed connections, respectively. In particular, two phenomena were found to be typical of the hysteretic response of steel dowel-type connections, as recalled in Ref. [43]. The first one is the *pinching effect* implying different hysteretic curves from the first to the subsequent load cycles (see **Figure 9**). The second one, referred to as the *memory of material*, is due to a dependence of the load-slip curve from the loading history. Both these phenomena may influence the ductile behavior of a timber structure.

3.1.1. Influence of the pinching effect on the ductile behavior of connections

The *pinching effect* is a very typical feature of the hysteretic behavior of dowel-type connections affecting both historical and modern timber constructions. The mechanical causes of it have been discussed in Section 2.2.1. This effect has been documented by many authors, as for instance [48–52]. It was found, in particular, that for a given displacement level, the highest resistance and widest hysteresis loop was attained at the first load cycle, whilst the subsequent cycles were narrowed and achieved lower resistance, stabilizing after about three cycles (see **Figures 9a** and **9b**). Stabilization of the pinched curve after three cycles is also referred to in UNI EN 12512:2006 [30]. Due to the reduction of the area of the hysteresis loop, the *pinching effect* may be actually responsible for a reduced amount of energy dissipation, although connections are still able to exhibit high values of ductility.

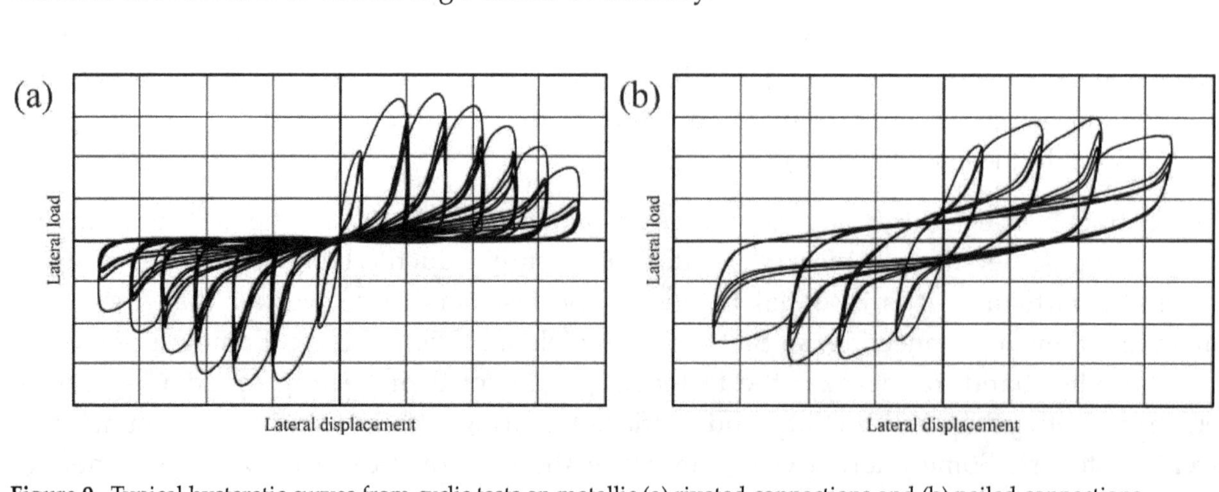

Figure 9. Typical hysteretic curves from cyclic tests on metallic (a) riveted connections and (b) nailed connections.

In modeling the mechanical behavior of a steel dowel-type connection for numerical analysis purposes, the pinching effect should be considered. A discussion on how this can be done may be found in Ref. [34], even if standard models comprehensive of the pinching effect and of the strength and stiffness degradation are not yet available, neither suggested by codes of practice.

3.1.2. Influence of the load history on the ductile behavior of connections

From the results available in the literature, it is clear that the hysteretic behavior of timber connections may strongly depend on the type of experimental test carried out (dynamic, static, cyclic, monotonic) as well as on the testing protocol adopted. On the other hand, while different protocols are available to carry out cyclic loading tests on timber structures, for example, EN 12512 [30], CUREE-Caltech standard [33], UBC protocol [11], a consensus on the best protocol to be assumed as standard has not been yet reached [48]. Many experimental findings proved, however, the influence of the load history on the final results.

It was shown in Ref. [48] that a connection usually reaches its maximum load at a lower deformation under cyclic loads than under monotonic loading. In Ref. [50], it was found that the ductility ratio of timber shear walls may be much higher when measured under static monotonic tests than when measured under dynamic tests. These experimental findings indicate that results from monotonic tests tend to overestimate the load-deformation behavior of connections with respect to cyclic loading tests, and therefore should be avoided when determining the seismic performance of timber buildings [48]. Dynamic tests are surely the best choice to capture the behavior of timber structures under seismic or wind loads, also in view of the fact that the failure modes may be very different in static and dynamic conditions [50]. The hysteresis loops obtained from dynamic tests were, however, found to be very sensitive to the protocol adopted [11, 53].

The dependence of the connection ductility from the experimental test may be also inferred from **Table 3**, where are collected the ductility ratios experimentally obtained for different timber connections [44, 48, 51–52, 54]. **Table 3** may be quite convenient to have an idea of the ductile capacity of timber connections, although the data herein provided should be compared with care, owing to the different specimens, test setups and loading protocols involved in the tests (the reader is referred to the papers quoted in the table for any detail).

Analogously, the ductility ratios of modern timber walls are given in **Table 4**, as derived from Refs. [50, 55, 56]. The data collected in **Table 4** highlight the good ductility which may be exhibited by modern timber constructions, although comparing the data collected in **Table 4** needs again caution. It can be also noted, finally, that the hysteresis curves obtained by testing modern timber walls with nailed connections evidenced features similar to those of **Figure 7**, as can be inferred for instance from diagrams provided in Ref. [50–51, 55, 57].

3.2. Non-linear dynamic analysis to predict the seismic response of timber structures

The non-linear time-history analysis (NLTHA) is the most comprehensive procedure allowed by seismic codes to design earthquake-resistant structures. It involves a complete time-history investigation under different spectrum-compatible ground motions. Despite its potential, the

Connection type	Wood elements	Loading	μ
Steel plates with bolts [48]	Glulam members	Monotonic	3–4.8
		Cyclic	2.53–2.91
Steel plates with glulam rivets [48]	Glulam members	Monotonic	16.4–20.4
		Cyclic	10.74–15.96
Steel brackets with nails or screws [51, 52]	XLam panels	Cyclic (parallel to grain)	3.01–6.36
		Cyclic (perpendicular to grain)	3.82–4.83
Dowel-type [44]	XLam members	Cyclic	1.3–2.1
Dowel-type reinforced with self-tapping screws [44]		Cyclic	3.4–7.3
Slotted-in steel plates with nails [54]	glulam members	Monotonic (parallel to grain)	11.9–31.9

Note: XLam, cross-laminated.

Table 3. Ductility of connections as obtained from experimental tests.

Test specimens	Connections	Loading	μ
Shear walls sheathed with plywood [50]	Plates to stud nailing	Monotonic	14
		cyclic	9.3
Shear walls sheathed with OSB [50]	Plates to stud nailing	Monotonic	13.2
		cyclic	7.7
Cross-laminated walls [55]	Hold-downs and brackets with nails, screws and rivets	Cyclic	3.65–7.54
Shear walls sheathed with OSB [56]	Nailed steel brackets and hold-down	Monotonic	3.5–4.9
		cyclic	3–4.2
Shear walls sheathed with GF [56]	Nailed steel brackets and hold-down	Cyclic	3.4
Shear walls sheathed with OSB and GF [56]	Nailed steel brackets and hold-down	Monotonic	5.67

Note: OSB, oriented strand board; GF, gypsum fiber.

Table 4. Ductility of modern timber walls as obtained from experimental tests.

NLTHA is still underused, likely due to the difficulties it indubitably involves and even to some inadequacies of the current codes of practice [58]. Such an analysis is, however, the better way to predict the actual seismic performance of structures composed of elastic and inelastic parts. Current codes of practice allow non-linear analyses for the calculation of the internal forces in the members of timber structures, provided that they are able to redistribute the internal forces via connections of adequate ductility [46].

When implementing a NLTHA, an efficient approach to model the structure is that of separating the critical zones where the NLTHA ductile behavior may be exhibited from the other structural parts which are expected to deform elastically even at the ultimate state. This is a typical procedure followed, for example, in reinforced concrete frames where plastic hinges are usually lumped at either end of columns and beams, while preventive plasticization of beams is guaranteed by some code-based strength hierarchy rules. An analogous procedure can be exploited for timber structures, by assuming timber members as purely elastic elements and connections as nonlinear links. To comply with the modern philosophy of the capacity design, the timber elements should be overdesigned to ensure their brittle failure will follow plasticization of connections (strength hierarchy rule).

3.2.1. Modeling timber connections

Exploiting experimental data is often the best way to obtain the mechanical behavior of a timber connection under dynamic loads. Several empirical models were proposed in the literature, which commonly involve parameters calibrated to experimental data, see for example [34, 43, 59, 60]. It should be noted, however, that extracting a general model from the experimental load-displacement curves needs caution owing to the possible dependence on both the loading history and the test set-up [34, 61, 62], as already discussed in Section 3.1.2. More detailed micro-models were also proposed by other authors, for example [62–64], which investigated the non-linear response of metallic fasteners and surrounding wood through three-dimensional finite element dynamic analyses. Still requiring some empirical adjustments of parameters, such sophisticated models usually imply a significant aggravation of the computational effort, which may become unsustainable for purposes other than those of advanced researches.

As already observed in Section 3.1, the behavior of timber connections depends from several factors, some of which are not easily predictable. This makes rather difficult to develop an analytical model able to reproduce the behavior of a timber connection. However difficult it may be, finding a suitable model for the hysteretic behavior of connections is essential to study the dynamic response of a timber structure, at least when a non-linear analysis has to be performed.

Commercial packages for structural analysis usually allow choosing between different mechanical models to implement the behavior of nonlinear links. For instance, the pivot hysteretic model provided by the widely used SAP2000 for nonlinear links (NLLINK) is depicted in **Figure 10**. To adopt a model like this, a set of parameters have to be properly assigned to reproduce all the typical phenomena experimentally detected in timber connections such as stiffness and strength degradation as well as pinching effects.

3.3. Concluding notes on the ductile behavior of modern timber structures

When appraising the ductile behavior of modern timber structures, the following aspects can be eventually addressed.

i. In-situ inspections, laboratory investigations and analytical models available in the literature converge to the statement that modern timber structures may be ductile and dissipative, provided that their metallic connections are well designed and detailed. Data collected

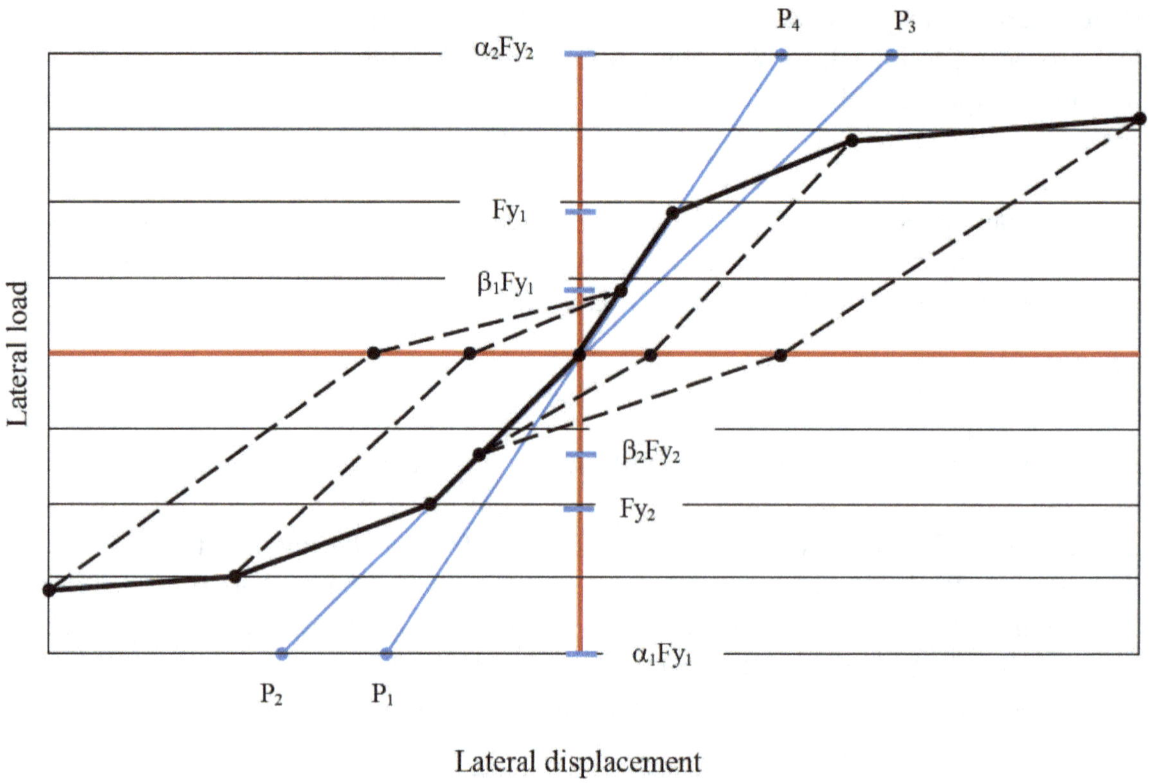

Figure 10. Multi-linear plastic pivot model for nonlinear links (NLLINK) in SAP2000.

in **Tables 3** and **4** show that timber constructions own a rather high—and sometimes excellent-ductility capacity. However, the actual ductility of a timber construction is not always univocally established. Different definitions of the ductility ratio are, in fact, proposed by authors and by codes of practice, as extensively discussed in Refs. [65–67].

ii. A decisive step to evaluate the ductility ratio, in its more classical definition, is the determination of both the deformation at yield and the ultimate deformation from the experimental load-deformation curve. Different conventional methods are available to this purpose as discussed in Refs. [66, 67], but adopting one method or another may lead to substantial differences in the calculation of the ductility ratio (even up to 100%), as evidenced in Ref. [66].

iii. The value of the ductility ratio alone cannot be sufficiently representative of the performance under strong dynamic loads. Not always, in fact, to higher values of ductility corresponds a greater amount of energy dissipated. This was addressed, for instance, in Ref. [55].

iv. To carry out a NLTHA of a timber structure nonlinear links should be located where plasticization is expected to occur, that is, at connections (plastic hinges). The mechanical model of the plastic hinges is generally difficult to be defined by the designer due to (a) insufficient support given by codes of practice; (b) difficulty in finding the experimental data relevant to the specific connection adopted; (c) uncertainty in the correct extraction of the required parameters from the experimental curves. These difficulties may eventually discourage the practitioners to carry out non-linear dynamic analyses of timber structures.

Author details

Maria Cristina Porcu

Address all correspondence to: mcporcu@unica.it

University of Cagliari, Cagliari, Italy

References

[1] CEN. Eurocode 8: Design of Structures for Earthquake Resistance – Part 1: General Rules, Seismic Actions and Rules for Buildings. European Comitee for Standardization, Brussels, 2004, 232 pp.

[2] Federal Emergency Management Agency (FEMA). NEHRP Recommended Seismic Provisions for New Buildings and Other Structures, FEMA P-750, Building Seismic Safety Commission, Federal Emergency Management Agency, Washington, D.C., 2009.

[3] American Society of Civil Engineers (ASCE). Seismic Evaluation of Existing Buildings, ASCE Standard ASCE/SEI 31-03, American Society of Civil Engineers/Structural Engineering Institute, Reston, VA, 2003.

[4] Porcu M.C. Reducing seismic stress on buildings through inertia limiters at floor level. In: Earthquake Resistant Engineering Structures IX. WIT Transactions on the Built Environment, vol. 120, WIT Press, UK, 2013.

[5] Porcu M.C. Plastic disconnection of storey masses for seismic stress control. WIT Transactions on the Built Environment. 2015;**152**:125-133. DOI: 10.2495/ERES150101

[6] Doğangün A., Tuluk Ö.İ., Livaoğlu R., Acar R. Traditional wooden buildings and their damages during earthquakes in Turkey. Engineering Failure Analysis. 2006;**13**(6):981-996. DOI:10.1016/j.engfailanal.2005.04.011

[7] Liska A.J., Bohannan B. Performance of wood construction in disaster areas. Journal of Structural Division, ASCE. 1973;**99**(12):2345-2354.

[8] Soltis L.A. Low-rise timber buildings subjected to seismic, wind, and snow loads. Journal of Structural Engineering. 1984;**110**(4):744-753. DOI: 10.1061/(ASCE)0733-9445(1984)110:4(744)

[9] Güçhan N.S. Observations on earthquake resistance of traditional timber-framed houses in Turkey. Building and Environment. 2007;**42**(2):840-851. DOI: 10.1016/j.buildenv.2005.09.027

[10] Rainer J.H., Karacabeyli E. Performance of wood-frame construction in earthquakes. Proceedings of the 12th World Conference on Earthquake Engineering, Auckland, New Zealand. vol. 30, 2000, **p**. 1-8.

[11] Lam F., Filiatrault A., Kawai N., Nakajima S., Yamaguchi N. Performance of timber buildings under seismic load. Part 1: Experimental studies. Progress in Structural Engineering and Materials. 2004;**4**(3):276-285. DOI: 10.1002/pse.121.

[12] Vivenzio G. Istoria e teoria de' tremuoti in generale ed in particolare di quelli della Calabria e di Messina del MDCCLXXXIII (History and theory of earthquakes in general and in particular of those occurred in Calabria and Messina in 1733). Stamperia Regale (Royal Printing House), Naples, Italy. 1783 (in Italian).

[13] Smith W.R. Building homes to withstand hurricane damage. Forest Products Journal. 1961;**11**(6):176-177.

[14] Vivenzio G. Istoria e teoria de' tremuoti in generale ed in particolare di quelli della Calabria, e di Messina del MDCCLXXXIII (History and theory of earthquakes in general and in particular of those occurred in Calabria and Messina in 1733). Stamperia Regale (Royal Printing House), Naples, Italy. 1783 (in Italian).

[15] Hamilton G. Relazione dell'ultimo terremoto delle Calabrie e della Sicilia inviata alla Società Reale di Londra (Report on the last earthquake of Calabria and Sicily sent to the Royal Society of London). Stamperia della Rovere ("della Rovere" Printing House), Florence, Italy; 1783 (in Italian).

[16] Ruggieri N., Tampone G., Zinno R. In-plane versus out-of-plane "Behavior" of an Italian timber framed system—The Borbone constructive system: Historical analysis and experimental evaluation. International Journal of Architectural Heritage. 2015;**9**(6):696-711. DOI: 10.1080/15583058.2015.1041189

[17] Ruggieri N., Zinno R. Mechanical and constructive interpretation of the giovanni vivenzio's model. In: Ruggieri N., Tampone G., Zinno R., editors. Historical Earthquake-Resistant Timber Frames in the Mediterranean Area. Switzerland: Springer International Publishing; 2015. p. 11-20. DOI: 10.1007/978-3-319-16187-7_2

[18] Mascarenhas J. Sistemas de Construção V: O Edifício de rendimento da baixa pombalina de Lisboa. Materiais Básicos (Construction systems: The building of income of the Pombaline downtown of Lisbon. Basic materials.). Ed. Livros Horizonte, Lisbon, 2005;3 (in Portuguese).

[19] Meireles H., Bento R., Cattari S., Lagomarsino S. A hysteretic model for "frontal" walls in Pombalino buildings. Bulletin of Earthquake Engineering. 2012;**10**(5):1481-1502. DOI: 10.1007/s10518-012-9360-0

[20] Mendes N., Lourenço P.B. Seismic assessment of masonry "Gaioleiro" buildings in Lisbon, Portugal. Journal of Earthquake Engineering. 2009;**14**(1):80-101. DOI: 10.1080/13632460902977474

[21] Aktas Y.D., Akyüz U., Türer A., Erdil B., Güçhan N.S. Seismic resistance evaluation of traditional ottoman timber-frame Hımış houses: Frame loadings and material tests. Earthquake Spectra. 2014;**30**(4):1711-1732. DOI: http://dx.doi.org/10.1193/011412EQS011M

[22] Dar M.A., Ahmad S. Traditional earthquake resistant systems of Kashmir. International Journal of Civil and Structural Engineering Research. 2015;**2**(2):86-92.

[23] Ali Q., Schacher T., Ashraf M., Alam B., Naeem A., Ahmad N., Umar M. In-plane behavior of the dhajji-dewari structural system (Wooden Braced Frame with Masonry Infill). Earthquake Spectra. 2012;**28**(3):835-858. DOI: http://dx.doi.org/10.1193/1.4000051

[24] Benson T. Building the Timber Frame House: The Revival of a Forgotten Craft. New York: Simon and Schuster; 1981.

[25] Qu Z., Dutu A., Zhong J., Sun J. Seismic Damage to Masonry-Infilled Timber Houses in the 2013 M7. 0 Lushan, China. Earthquake Spectra. 2015;**31**(3):1859-1874. DOI: http://dx.doi.org/10.1193/012914EQS023T.

[26] Quinn N., D'Ayala D. In-plane experimental testing on historic quincha walls. In: 9th International Conference on Structural Analysis of Historical Constructions; F. Peña & M. Chávez (eds.), Mexico City, Mexico, 2014.

[27] Cruz H., Moura J.P., Machado J.S. The use of FRP in the strengthening of timber reinforced masonry load-bearing walls. Historical constructions – Guimarães, Portugal. 2001, p. 847-856.

[28] Santos P. Ensaios de paredes "pombalinas" (Tests on "pombalinos" walls). Technical note No. 15/97. 1997; NCE/DE, 17 LNEC Lisbon (in Portuguese).

[29] D'Ayala D., Speranza E. An integrated procedure for the assessment of seismic vulnerability of historic buildings. In: Proceedings of the 12th European Conference on Earthquake Engineering; London, UK: Elsevier Science Limited; 2003. Paper Reference 561.

[30] UNI EN 12512. 2006. Timber structures. Test methods. Cyclic testing of joints made with mechanical fasteners. UNI: Milano, Italy.

[31] Jacobsen L.S. Steady forced vibrations as influenced by damping. Transactions of ASME, 1930;**52**(15):169-181.

[32] Vasconcelos G., Poletti E., Salavessa E., Jesus A.M., Lourenço P.B., Pilaon P. In-plane shear behaviour of traditional timber walls. Engineering Structures. 2013;**56**:1028-1048. DOI: 10.1016/j.engstruct.2013.05.017

[33] Krawinkler H., Parisi F., Ibarra L., Ayoub A., Medina R. Development of a Testing Protocol for Wood frame Structures. California: CUREE Publication; 2000;W-02:1-87.

[34] Foliente G.C. Hysteresis modeling of wood joints and structural systems. Journal of Structural Engineering. 1995;**121**(6):1013-1022. DOI: 10.1061/(ASCE)0733-9445(1995)121:6(1013)

[35] Fang D.P., Iwasaki S., Yu M.H., Shen Q.P., Miyamoto Y., Hikosaka H. Ancient Chinese timber architecture. I: Experimental study. Journal of Structural Engineering. 2001;**127**(11):1348-1357. DOI: 10.1061/(ASCE)0733-9445(2001)127:11(1348)

[36] Fang D.P., Iwasaki S., Yu M.H., Shen Q.P., Miyamoto Y., Hikosaka H. Ancient Chinese timber architecture. II: Dynamic characteristics. Journal of Structural Engineering. 2001;**127**(11):1358-1364. DOI: 10.1061/(ASCE)0733-9445(2001)127:11(1358)

[37] Li X., Zhao J., Ma G., Chen W. Experimental study on the seismic performance of a double-span traditional timber frame. Engineering Structures. 2015;**48**:141-150. DOI: 10.1016/j.engstruct.2015.04.031

[38] Maeno M., Suzuki Y., Ohshita T., Kitahara A. Seismic response characteristics of traditional wooden frame by full-scale dynamic and static tests. In: 13th World Conference on Earthquake Engineering, Vancouver, Canada, 2004

[39] Suzuki Y., Maeno M. Structural mechanism of traditional wooden frames by dynamic and static tests. Structural Control and Health Monitoring. 2006;**13**(1):508-522. DOI: 10.1002/stc.153

[40] Tanahashi H., Suzuki Y. Basic concept and general formulation of restoring force characteristics of traditional wooden joints. In: Proceedings of the 13th World Conference on Timber Engineering; Auckland. 2012. p. 378-387.

[41] Makarios T., Demosthenous M. Seismic response of traditional buildings of Lefkas Island, Greece. Engineering Structures. 2006;**28**(2):264-278. DOI: http://dx.doi.org/10.1016/j.engstruct.2005.08.002

[42] Parisi M. A., Piazza M. Seismic behavior and retrofitting of joints in traditional timber roof structures. Soil Dynamics and Earthquake Engineering. 2002;**22**(9):1183-1191. DOI: 10.1016/S0267-7261(02)00146-X.

[43] Loss C., Piazza M., Zonta D. Direct displacement-based seismic design of timber structures with dowel-type fastener connections. In: World Conference on Earthquake Engineering, Lisboa, 2015.

[44] Blaß H.J., Schädle P. Ductility aspects of reinforced and non-reinforced timber joints. Engineering Structures. 2011;**33**(11):3018-3026. DOI: http://dx.doi.org/10.1016/j.engstruct.2011.02.001

[45] Quenneville J.H., Mohammad M. On the failure modes and strength of steel-wood-steel bolted timber connections loaded parallel-to-grain. Canadian Journal of Civil Engineering. 2000;**27**(4):761-773. DOI: 10.1139/l00-020

[46] EN 1995-1-2:2004. Eurocode 5: Design of timber structures-Part 1-1: General – Common 29 rules and rules for buildings. CEN, Brussels, 2004.

[47] Rodd P.D., Leijten A.J.M. High-performance dowel-type joints for timber structures. Progress in Structural Engineering and Materials. 2003;**5**(2):77-89. DOI: 10.1002/pse.144

[48] Popovski M., Prion H.G., Karacabeyli E. Seismic performance of connections in heavy timber construction. Canadian Journal of Civil Engineering. 2002;**29**(3):389-399. DOI: 10.1139/l02-020

[49] Popovski M., Karacabeyli E. Seismic performance of riveted connections in heavy timber construction. In: Proceedings of the 13th World Conference on Earthquake Engineering, Vancouver, Canada, 2004. Paper No. 3356.

[50] Dinehart D.W., Shenton III H.W. Comparison of static and dynamic response of timber shear walls. Journal of Structural Engineering. 1998;**124**(6):686-695. DOI: 10.1061/(ASCE)0733-9445(1998)124:6(686)

[51] Schneider J., Karacabeyli E., Popovski M., Stiemer S.F., Tesfamariam S. Damage assessment of connections used in cross-laminated timber subject to cyclic loads. Journal of Performance of Constructed Facilities. 2013;**28**(6):A4014008. DOI: 10.1061/(ASCE)CF.1943-5509.0000528

[52] Shen Y.L., Schneider J., Tesfamariam S., Stiemer S.F., Mu Z.G. Hysteresis behavior of bracket connection in cross-laminated-timber shear walls. Construction and Building Materials. 2013;**48**:980-991. DOI: http://dx.doi.org/10.1016/j.conbuildmat.2013.07.050

[53] Gatto K., Uang C.M. Effects of loading protocol on the cyclic response of wood frame shear walls. Journal of Structural Engineering. 2003;**129**(10):1384-1393. DOI: 10.1061/(ASCE)0733-9445(2003)129:10(1384)

[54] Stehn L., Johansson H. Ductility aspects in nailed glue laminated timber connection design. Journal of Structural Engineering. 2002;**128**(3):382-389. DOI: 10.1061/(ASCE)0733-9445(2002)128:3(382)

[55] Gavric I., Fragiacomo M., Popovski M., Ceccotti A. Behaviour of cross-laminated timber panels under cyclic loads. In: Materials and Joints in Timber Structures. Springer Netherlands, 2014. p. 689-702. DOI: 10.1007/978-94-007-7811-5_62

[56] Sartori T., Piazza M., Tomasi R., Grossi P. Characterization of the mechanical behaviour of light-frame timber shear walls through full-scale tests. In: World Conference on Timber Engineering, 2012, July, p. 180-188.

[57] Ibarra L.F., Medina R.A., Krawinkler H. Hysteretic models that incorporate strength and stiffness deterioration. Earthquake Engineering &Structural Dynamics. 2005;**34**(12):1489-1511. DOI: 10.1002/eqe.495

[58] Porcu M.C. Code inadequacies discouraging the non-linear dynamic analysis of buildings. In: Earthquake Resistant Engineering Structures XI, Alicante, Spain, WIT Press, 2017.

[59] Baber T.T., Noori M.N. Random vibration of degrading, pinching systems. Journal of Engineering Mechanics. 1985;**111**(8):1010-1026. DOI: 10.1061/(ASCE)0733-9399(1985)111:8(1010)

[60] Dowrick, D. J. Hysteresis loops for timber structures. Bulletin of New Zealand National Society of Earthquake Engineering. 1986;**19**(20):143-152.

[61] He M., Lam F., Prion H.G. Influence of cyclic test protocols on performance of wood-based shear walls. Canadian Journal of Civil Engineering. 1998;**25**(3):539-550.

[62] Foschi, R.O. Modeling the hysteretic response of mechanical connections for wood structures. In: World Conference of Timber Engineering, Whistler, BC, Canada. 2000, July.

[63] Lo, Y.P. A comparison of two hysteretic models in predicting the response of a connector under cyclic loading (Doctoral dissertation, University of British Columbia), 2002.

[64] Lam F., Filiatrault A., Kawai N., Nakajima S., Yamaguchi N. Performance of timber buildings under seismic load. Part 2: Modelling. Progress in Structural Engineering and Materials. 2004;**6**(2):79-83. DOI: 10.1002/pse.175

[65] Jorissen A., Fragiacomo M. General notes on ductility in timber structures. Engineering Structures. 2011;**33**(11):2987-2997. DOI: http://dx.doi.org/10.1016/j.engstruct.2011.07.024

[66] Munoz, W., Salenikovich, A., Mohammad, M., & Quenneville, P. Determination of yield point and ductility of timber assemblies: in search for a harmonized approach, Proc. of Meeting 41 of CIB-W18, Canada, St Andrews, 2008.

[67] Brühl F., Kuhlmann U., Jorissen A. Consideration of plasticity within the design of timber structures due to connection ductility. Engineering Structures. 2011;**33**(11):3007-3017. DOI: http://dx.doi.org/10.1016/j.engstruct.2011.08.013

4

Wood-Reinforced Polymer Composites

Anil Akdogan and Ali Serdar Vanli

Abstract

The importance of protection of natural resources, recycling and biodegrability are the basic reasons behind the increased use of lignocellulosic (LC) material–reinforced polymer composites. This chapter investigates wood filler–reinforced polymer composites, which have gained attention among the researchers and in the market in woody applications having aesthetic and low service costs. Most polymers have the increasing cost of petrochemical products while LC materials are obtained from different national resources such as different types of wood powder. LC materials are available in light, cheap, environment-friendly forms with easy maintenance. The main disadvantages of LC-reinforced composites are moisture sensitivity and poor dimensional stability, besides changing wood fiber characteristics in time and poor adhesion to basically all kind of matrix polymers. Additive usage in manufacturing is an increasing trend of wood-based LC material–reinforced polymer composite parts to penetrate new markets, with better performance requirements and long-term stability. This chapter investigates description, classification, production with common additive types and usage of wood-polymer composite materials. In addition, it approaches preprocesses to reach high productivity and high-quality product manufacturing. Finally, some experimental manufacturing results of wood-reinforced polymer composites are presented.

Keywords: polymer composites, wood reinforcements, lignocellulosic materials, manufacturing methods, product quality

1. Introduction

Using lignocellulosic (LC) materials as filler has valued attention in the polymer composite production industry. While LC materials are obtained from different national resources like different types of wood powder, polymers have the increasing cost of petrochemical products [1]. The trend was provided both by exchanging petrochemical materials for increased bioderived sustainable materials and the need to reduce waste to landfill and explore recycling technologies. Wood polymer composite (WPC) products having wastepaper and/or second hand plastics into are some of the major components of global municipal solid waste which can be recycled [2, 3]. Engineers and scientists have been working to improve the various qualifications of both thermosets and thermoplastics through the addition of wood fiber since the beginning of the twentieth century. The concept is still open to improve and seems will be for a long time. Fine powder of wood or woody biomass from agricultural residues is mixed with polymers such as polyethylene or polypropylene generally within an adjusted extruder. The extruded material can be pelletized for final processing of products by injection molding, extrusion or compression molding, or the extruded material may be formed into simple sections such as planks or more complex profiles for specific applications. LC filler might be obtained from different national resources such as nut shells and residual wood powder. The well-known form of wood filler for thermoplastics is wood powder. Wood powder can be defined as a finely granulated form of wood. It is derived from various wood shavings, sawdust, chips and other waste wood from saw mills and residues of other wood processing industries. Wood powder is available commercially in a variety of size distributions and species in the market [4].

LC materials are available in light, cheap, environment friendly forms with easy maintenance. Having lower cost per unit volume of wood fillers, low wear effects on processing machines and equipment, lack of health hazards, lower densities, abilities for surface modifications and their abundance in nature make them suitable solution in both filler and fiber forms compared to other inorganic fillers. The well-known handicaps of LC-reinforced composites are their moisture sensitivity and poor dimensional stability, besides changing wood fiber characteristics in time with origin and the time of the harvest and poor adhesion to basically all kind of matrix polymers. The main application areas of LC polymer composites are building, automotive, furniture and packaging industries besides flooring, fencing, trim board, windows, plywood, roofing and siding. Other applications are in infrastructure, e.g., boardwalks, marinas and guardrails; in transportation, e.g., interior automotive panels and dashboards, truck floors and head liners; and in industrial and consumer applications, e.g., playground equipment and pallets. Additionally, outdoor tables chairs, crossovers, sun shades and benches are already being made of wood polymer composites. The WPC can be used in food packaging industry as an alternative biodegradable product in the case of high percentage LC filler reinforcement [5–8]. The range of fibers and fillers are typically introduced at loadings, i.e., weight percentage, of 30–70% in the polymer, although laboratory studies reporting levels from 5% upward exist. Although the situation in tensile properties is more variable, continued improvement could be achieved in bending properties at high filler loadings [2]. Plastic as matrix and wood powder as filler constitute the main ingredients in WPC receipts. The weight percentage of wood powder varies in the receipts and wood powder content of up to 70% is currently used in commercial WPC products [4].

Products from WPCs have been manufacturing in industry for a long time. In practice, using polymer processing machines to produce WPC bring some important processing problems causing high production costs [5]. In order to overcome the hydrophilic nature of wood which is a big handicap and to improve the adhesion and compatibility of polymers to the cellulose of wood, researchers use preprocessing for recent years [7–9]. Much research has been centered on aspects of compatibility between the commonly used plastic matrix and the wood fiber or powder to accelerate WPCs performance [5, 6, 9–11]. Other efforts have been made to reduce the density, by foaming of the manufactured profile or by designing hollow sectioned parts for specific applications [2, 12]. WPCs are in general decrease material costs and increase stiffness, basically provided by lignin, compared to the virgin polymers at the expense of ductility and impact strength. In addition, these composites could be manufactured by existing cost-effective processing equipment with a few but important adjustments. It is therefore important to review the literature on wood filler material–reinforced polymer composites. In the literature, there are many works on production and characterization of various kinds of LC fillers and polymers prescribing valuable composite receipts [6, 7, 13–15].

Some researchers processed composites with wood powder and wood chips as filler and various PP grades as the matrix polymer. The measurement range for filler particle lengths and widths are in a wide range. Tensile strength, elongation at break and impact strength were significantly increased in composites reinforced with wood fibers, upon introduction of coupling agents. Scanning Electron Microscope (SEM) examinations revealed well-dispersed wood powders in the extruded compounds and injection molded test samples. The dispersion of the wood fiber quality was in general significantly improved by introduction of coupling agents [16]. In other work, researchers focused on extrusion and injection molding of different biocomposites, containing up to 70 wt.% wood powder. In general, at higher filling levels, lower impacts of fiber content on mechanical properties were determined. The tensile and flexural strengths of all biopolymers were significantly improved except the poly-lactic acid. About 180% increase in tensile strength was achieved with the aliphatic-aromatic co-polyester produced from fossil resources at maximum fiber load of 65 wt.% [6]. In another research work, using wood particle filler for WPC production suggest that processing of composites having wood content above 60% is highly challenging, especially where the wood particles are fine [14]. However, some researchers measured the length and diameter of the cellulose fibers to study how the different processing steps are affected by the dimensions of the fibers and their reinforcing capacity.

Additionally, the authors of this chapter made different investigations on the effects of three different dimensioned beech powder as additives to the morphological and mechanical properties of the polymer composite materials. At lower particle sizes of certain concentrations of wood powder loads result in better mechanical properties and morphological improvements of composites are experienced with a sharp decrease in the elongation at break. The scanning electron microscope (SEM) analysis (**Figure 1**) at fractured surfaces of tensile bars indicate that smaller filler sizes cause homogenize structures but if the fiber size is too small, the matrix cannot grip the fiber [5].

Moreover, there are some works considering wood fiber quality, fiber modification and especially thermal treatments on the physical and mechanical properties of WPCs. Tensile mechanical investigations show an increase in rigidity of the composites upon increasing wood powder content, together with a sharp decrease in the elongation at break. There were often slightly better results with the samples which had undergone a drying pretreatment [6, 9–11, 13, 14, 16]. Low resin ratio and viscosity, easy providing, low grain hardness and strength and also low cellular moisture features are desired for LC materials in order to use them in WPC products. Because of their available properties, it is possible to meet many of the herbal materials such as beech, pinewood, spruce, walnut, ailanthus, naturally desiccated grass pedicles and various kinds of plants, as composite material fillers in the literature [15, 17–20]. The basic reason of the growing usage of such natural fibers is their strong fiber structure. Producers are also penetrated into trim board and plywood markets using lightness, abrasiveness and environmental friendly specifications of the WPC materials [21, 22].

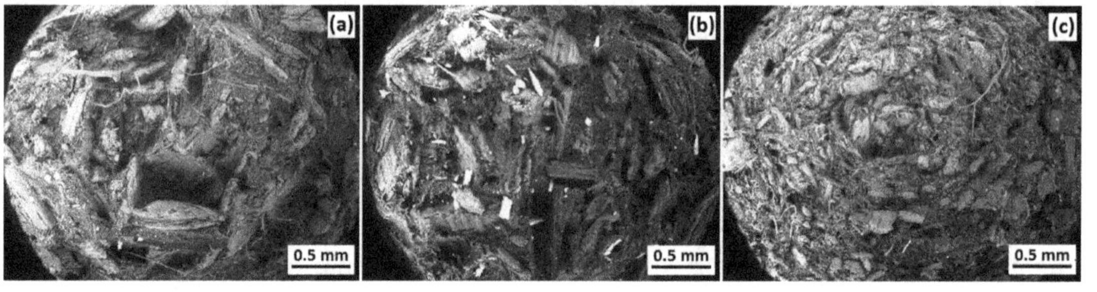

Figure 1. SEM micrographs of the fracture surfaces of the different size wood filler–reinforced composites (a) 1200–1800 μm sieve mesh sized wood filler–reinforced composite (b) 800–400μm sieve mesh sized wood filler–reinforced composite (c) under of 400μm sieve mesh sized wood filler–reinforced composite [5].

2. WPC compositions, types and preprocesses

WPCs are made up of a matrix polymer and wood material as fiber or a few kinds of filler forms and various kinds of processing agents or additives. This chapter gives a brief review of WPC compositions, types and preprocessing applications.

Most synthetic polymers are derived largely from petrochemicals. Others are derived from natural resins such as gutta-percha, tar, rosins and other modified natural sources such as cellulose/hemicelluloses and protein. Polymer materials having covalent bonds provide rigid and strong bonding of atoms within long-chain molecules. Polyethylene (PE) is accepted as the simplest polymer. Molecular weight of a polymer can be determined by multiplying the molecular weight of the mer by the degree of polymerization. The properties of polymers are formed by the degree of polymerization. Generally, low degrees of polymerization are achieved for surface finishes even coatings; whereas, high degrees of polymerization are necessary for the polymers shaped by extrusion or injection molding. Thermoplastic polymers have no cross-linking, so that they can be reshaped by heating to evaluated temperatures; whereas, thermosetting polymers having extensive cross-linking are rigid and cannot be reshaped when heated. It means the extension of cross-linking between the long-chain molecules is a method of distinguishing the commercially available polymers [22]. Although thermoset polymers have higher thermal stability and lower water absorption compared to thermoplastics, the increased demand for recyclability aims the researchers and users at thermoplastic-based composites. Most of the single polymer plastics made of petroleum are relatively easy to recycle [1, 23]. Commonly used manufacturing thermoplastic composites include polyethylene (PE) both high density (HD) and low density (LD) types, polypropylene (PP), polyvinyl chloride (PVC), polyvinyledene fluoride (PVdF), polybutylene terephthalate (PBT), polyvinyledene oxide (PPO), acrylonitrile-butadiene-styrene (ABS), acrylonitrile-styrene-acrylate (ASA), polycarbonate (PC) and some others [24, 25]. On the basis of T_m (melting temperature), the main three candidates are PE, PP and PVC. Their melting points are below 180°C and given in **Table 1** with some other important specifications like T_g (glass transition temperature) and densities. PE, PP, PVC including polystyrene (PS) are used in WPCs frequently because of their lower processing temperatures (150–220°C) than the others which prevent the degradation of the reinforcement materials [4].

Polymer	T_m (°C)	T_g (°C)	Density (g/cm³)
PE (LD)	98–115	−125	0.91–0.92
PE (HD)	120–137	−130 to 100	0.94–0.96
PP	160–173	−20 to 0	0.90–0.91
PVC	100–260	80 to 87	1.16–1.38

Table 1. Three basic polymers with specifications.

The primarily limiting factor of the use of thermoplastics in WPC production is the thermal degradation of wood fiber. Typically, wood starts to degrade around 220°C and this serves as the upper limit for the melting point of thermoplastic matrix resins. Filler content in WPC is generally in the range of 40–70% by weight. Flexural modulus, heat distortion temperature and impact strength are the basic mechanical properties besides important physical expectations like densities considered by the investigators and users [1].

The chemical composition of wood filler is quite complex. It varies according to the species. It is determined in LC material classification. It is made up of cellulose/hemicellulose, lignin and many other extractives. Firstly, the main structural component of wood is cellulose. The cellulose molecule is almost totally linear. Cellulose functions as the primary structural and the most abundant component within the wood fiber cell walls. Cellulose is hygroscopic and water absorption of it depends on the number of free hydroxyl groups. Since hemicellulose has a short chain with a degree of polymerization, it has low molecular weight. Cellulose serves as connecting agent that links bonds to the micro fibrils providing a kind of structural reinforcement to the wood fiber cell wall. The typical L/D (fiber length to diameter ratio) of wood powder ranges from about 3 to 5. Wood particles are generally defined by sieve mesh sizes. Secondly, lignin provides stiffness to wood by acting as cement in bonding the cellulose filaments. It is a brittle, an amorphous polymeric material, relatively inert that acts as both a bonding agent and a stiffening agent. Lignin can be thought of as a matrix that connects cellulose fibers to each other. They can be removed by some organic solvents or water. Finally, wood extractives are known as low molecular weight oleophilic compounds. Since they are organic substances, they can be removed by the related solvents. Extractives can include organic waxes, oils, fats, gums, carbohydrates, acids and resins at different percentages. Their common effects are hygroscopicity, permeability, long-term stability and durability of wood. Ether, alcohol or water extractions are the typical well-known extractives removed by the steam distillation [1–2, 4].

Preprocessing of the filler ranges from simple sizing/sieving to pulping and chemical modification of the surface, depending on the required quality specifications from the final application. Sizing of wood generally operates automatically at the designed shaking sieving machines. Since wood particles are generally defined by sieve mesh sizes, machines can include various screens. A typical sieving process can be supplied by an electric motor as automatically and consistently to feed the WPC mass production lines. As an example, three different sizes of mesh screens are placed to the designed shaking sieving machine to use in the experiments. The schematic drawing of the machine is given in **Figure 2** [26].

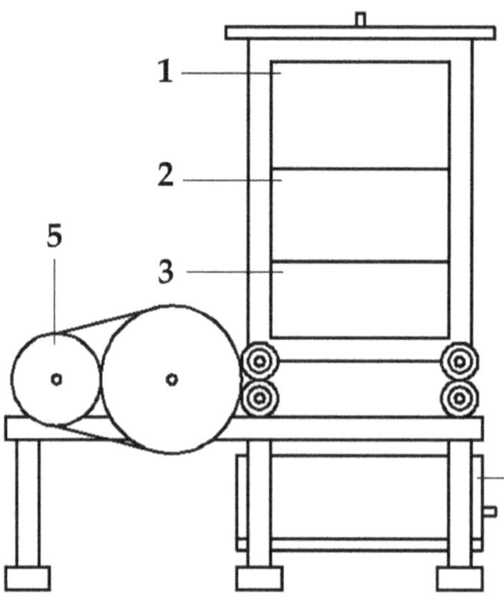

No	Description
1	1200 μm mesh screen
2	800 μm mesh screen
3	400 μm mesh screen
4	Under of 400 μm dust
5	Electrical motor

Figure 2. Shaking sieving machine [26].

Since interfacial adhesion between the polymer matrix and the wood filler defines the final physical properties of a WPC, in most WPC production operations, it is absolutely necessary to compatibilize or couple the mix. Various wood modification techniques used on solid wood to achieve dimensional stabilization besides polymer modification were investigated to assess the effect on bonding. In addition to the compatibilizers, techniques are frequently investigated which will alter the fiber surface chemistry or physical properties as a result of simple pretreatments [2]. In the literature for the modification of the wood fiber, much chemistry have been reacted with wood such as aldehydes, epoxides, anhydrides, siloxanes and isocyanates in an effort to compatabilize of wood for different polymer matrixes. It is clearly seen that as an alternative to the conventional pulping processes, solvent pulping method plays an important role [7, 10, 20, 27]. Subsequently, many investigators have reported studies on the reaction of wood with a large number of alcohols. For instance, in a research work, switchgrass was used in the pulping trials. A laboratory digester, which was equipped with a liquor recirculation system and external indirect electric heating element, was used in the cooking process. After cooking, the material was de-fiberized in a blender and washed on a filter paper. The washed material was dried. The soda pulp from switchgrass supplied excellent mechanical properties and showed a great potential as a reinforcement component [27]. On the other hand, wood chips were delignificated in glycerol solvent. At the end of delignification, wood washed with cold water and then with hot water until a clean filtrate was obtained. Due to its highly polar polyalcohol structure, glycerol could easily penetrate into the wood tissue. Thus, glycerol provides an effective reaction medium for delignification of wood chips. In addition, it was an excellent solvent for Na_2CO_3 and NaOH used as catalysts [28]. There are many other works pulping many kinds of LC materials or polymers at many different combinations of solvents [7, 15, 20, 29].

It is find out in the early researches that the interfacial bonding between the fiber and polyolefin matrix was very low. LC materials also would have high density of hydroxyl groups on their surface due to cellulose and hemicelluloses. Moreover, the lignin contains some polar functional groups. On the other hand, PE is a linear alkane of very high molecular weight, containing only repeating CH_2 units. The surface-free energy of PE is 31 mJ/m^2; whereas, 32 mJ/m^2 for PP [30] and although the surface energy of wood is much higher, 40–60 mJ/m^2, it typically contributes only as 31.5 mJ/m^2 to the surface energy of WPCs [31]. This difference in surface energy can lead to poor dispersion of the fiber within the matrix in blending. It even results in relatively weak interfacial bonding between the two phases of the WPC. A wide range of compounds were investigated to overcome these difficulties in many researches [2].

The need for coupling agents to improve the fiber-matrix adhesion in LC material-filled composites is well-documented. Various methods to improve interfacial adhesion have been examined. Enhanced interfacial adhesion and aid in fiber dispersion can be supplied by coupling agents. A notified increase in interfacial adhesion between the polymer and the fiber certainly results an improvement in mechanical and rheological properties of the composite aiding processing ability [1, 5]. Coupling agents become more critical growing product lifetime and strength requirements. Coupling agents used in WPCs play a vital role in adhesion between nonpolar polymeric matrix and polar wood fibers improving compatibility even in high filler loadings. Besides, increased filler loadings make the composites stiffer but reduce the toughness. Today, coupling agents are being used mostly in automotive and construction applications to improve long-term product performances. They help fiber dispersion, which results in reduced water absorption and better maintenance of mechanical properties after water exposure. Coupling agents increase flexural and tensile strengths and room temperature creep resistance. Moreover, heat deflection temperature increase by corresponding better creep resistance at elevated temperatures [32]. In general, wood filler improves the heat deflection of unfilled polymer matrix. Coupling agents can also aid in foaming by improving blend strength at the die exit and by conducting dispersion [33].

In the case of using polyolefins in WPC production, maleic anhydride-grafted polymers are widely being used in industry. Maleic anhydride grafted polypropylene (MAPE) and maleic anhydride grafted polyethylene (MAPP) appear like quite suitable agents, which are added to the blend. Small quantities of the MAPE or MAPP supply improvements in mechanical properties of the blend when added to the wood powder and polymer mix during compounding [8, 34–37]. There are a large number of brands and grades of MAPP are available and it is nowadays one of the most commonly used compatibilizer in the PP used WPC industry; whereas, MAPE is widely accepted for the PE-based WPCs. The other popular coupling agents are presently being used include anhydrides, isocyanates, silanes and other anhydride-modified copolymers in industry. More than forty coupling agents are being used not only in research works but also in manufacturing for a long time. These agents are grouped as organics and inorganics. Organic agents are preferred much more than inorganic agents due to their stronger interfacial adhesion. Coupling agents are generally coated on the surface of wood filler, polymer or both by compounding, blending, mixing, soaking, spraying, or other methods [12]. These agents are used for wood fiber and other fillers to improve reinforcing ability over the functionality of the fiber surfaces. Also, the weight fraction of the ingredients affects the final physical properties of the blend. Such as block copolymers and surfactants are commercially available compatibilization chemistries. But the common commercially available coupling agents include derivatives of maleic anhydride or siloxanes as reactive groups.

Processing additive usage in manufacturing is an increasing trend of wood-based LC material–reinforced polymer composite parts to penetrate new markets with better performance requirements and long-term stability. The use of colorants and lubricants as additives is well established besides the others like antimicrobials to maintain both appearance and physical integrity and ultraviolet stabilizers which protect the polymer portion of the composite from degradation. Antimicrobial additives may be used, especially for outside usage such as decking, sliding and living set products. The customer may be concerned colorants against microbial attacks. It may also be necessary to protect the biodegradable woody component of the WPC material from microbial attacks as a result of high humidity conditions. Zinc borate is known as the most popular antimicrobial additive. Lubricants are typically used for increasing output volume of many extrusion processes [32]. Processing additives are generally added for better blending and improving the smoothness of the WPC product while profile getting out of the die. This can avoid forming alligator-type skin textures which are the results of high shear forces. Lubricants reduce shear between the filler and the polymer matrix and aid fine blending of the WPC. Moreover, fatty acid esters and alcohols or metal soap-like zinc stearate may be used for lubrication. While metallic stearate lubricants can interfere with the maleic anhydride functionalized polyolefin-coupling agents used in some WPCs, alternative lubricant systems are available [38].

On the other hand, foaming of extruded or injection molded parts takes great attention. Foaming of WPCs is achieved by dispersing a gaseous phase throughout the liquid polymer and filler to create foam comprising a polymer and included gas component. The preservation of the resultant foamed state is important to maintaining the required structural properties. Foamed WPC parts can be manufactured using either chemical, generating a foaming gas by a chemical reaction or physical, achieving foaming with any chemical change, blowing agents. Among them, foaming of WPC is getting increasingly important day by day. The primary benefits of foaming are lightening weight of the compound, improving durability of the product and decreasing costs of the manufacturing. Moreover, foaming increases stiffness and makes the WPC more like wood for cutting and assembling type operations [32]. Foaming helps to improve performances of wood-working tools and fasteners by reducing part density and improving stiffness. It reduces cycling times that increases process performance and productivity. Reduced weight makes WPC aid to use in both current and new applications. Since amorphous polymers, e.g., polyvinyl chloride and polystyrene have high melt strength, they can be foamed well. Crystalline polymers such as PP and linear PE are more difficult to foam because they have low melt strength and typically contain calcium stearate, which acts as a de-foamer. That is why choosing a compatible lubricant is also crucial. Foaming agent performance is dependent on the ability to blend, pressure practice and temperature control [37]. Considering all components, the prepared flow chart in **Figure 3** would be helpful to summarize WPC to part manufacturing.

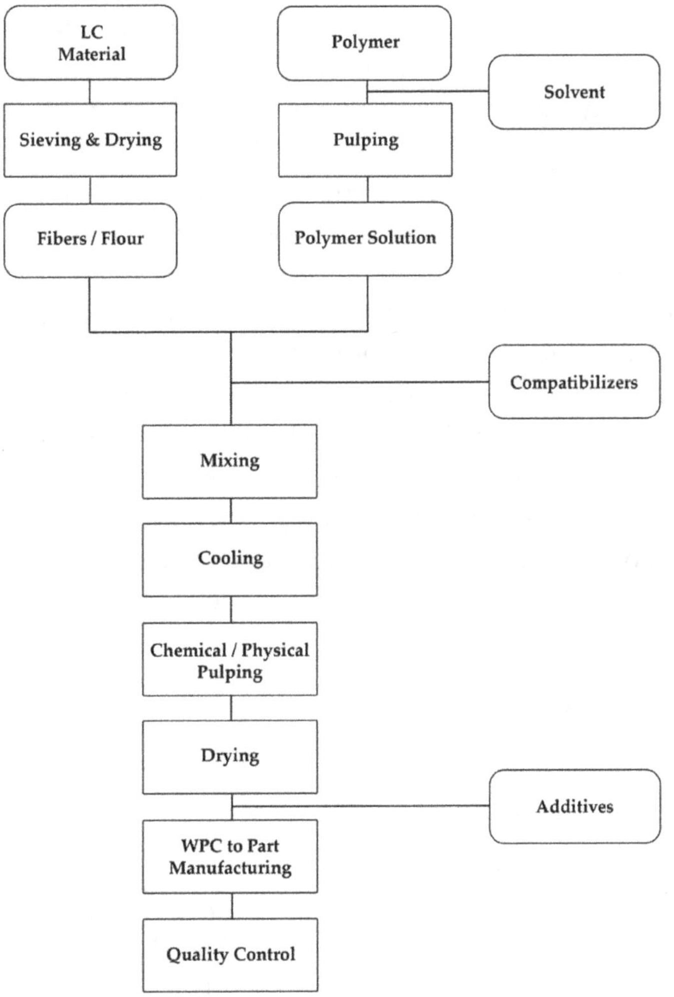

Figure 3. WPC to part manufacturing flow chart.

In addition, water absorption in finished parts is a basic problem for foamed WPC [32]. Control of moisture in the WPC production process is crucial because water anhydrides catalyze the polymer. The WPC material requires effective mixing, degassing and accurate metering processes and controlling the temperature. Of course, wood powder is very sensitive to moisture; it needs to be controlled. Moisture level of wood powder above 1% in the extruder causes production of micro-voids of irregular and inhomogeneous shapes which affect the usage performance of the products [39, 40]. Wood contains moisture which amount must be reduced in acceptable levels for WPC production. Over moisture in LC materials affects the composite production process negatively. Considering this truth, wood scrap should be dried efficiently and homogeneously in a short time. Also the moistening of additive materials in storage conditions should be determined so that available time period of manufacturing for dehumidified material must be stated. Every manufacturer should determine this period according to their individual manufacturing parameters, the used LC material properties and the storage conditions.

The moisture affinity of the filler materials are investigated in the experiments [5]. The wood scrap used in the experiments was dried in the designed fluidized bed drying machine by the authors. The machine has three essential stations for the wood material. The first station, the back funnel, is a small part acting as a silo for the material. It is shown in **Figure 4**, with the number 6. There is a decaying pitch screwed conveyor system under the back funnel. This system transfers the wood to the next station. Its rotational speed can be changed according to the desired raw material volume. The second station, the fluidized bed, is the main part of the machine acting as a pneumatic conveyor for the material. This station is shown with the number 8. The fan, indicated by the number 11, rises up the solid particles in the fluidized bed. Fluid air temperature could be increased by a heat-controlled resistance placed inside or outside of the fluidized bed so that the required dry air can be achieved. However, inside temperature of the fluidized bed must be lower than the decomposition temperature of wood material. Otherwise, undesired burnings and ash formations may occur. Excess air can be evacuated by the filter cloth, shown by the number 5, on the transmission channel. Then, the material is transferred to the pre-funnel indicated by the number 4. After that, the material reaches the last station of the system, the heating bed, shown with the number 3. The material is transferred into a glass flow channel by a stable pitch helical shaft, in which an infrared

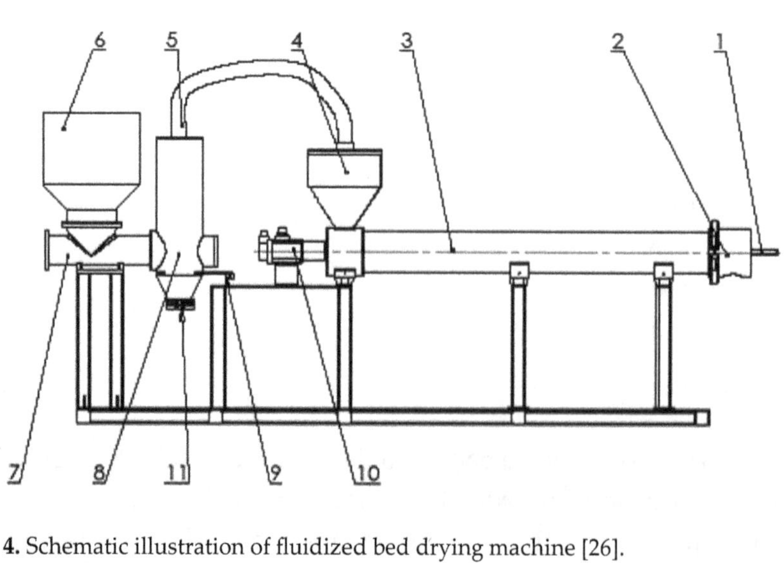

No	Description
1	Infrared Heater
2	Evacuation Flange
3	Heating Bed
4	Pre-Funnel
5	Filter Cloth
6	Back Funnel
7	Volumetric Dosing
8	Fluidized Bed
9	Single Track Valve
10	Motor
11	Fan

Figure 4. Schematic illustration of fluidized bed drying machine [26].

heater is fixed. Infrared heater temperature could be set digitally between 15 and 200°C. Nitrogen could be injected into the glass flow channel for the purpose of deporting excess moisture and foreign particles. The flow channel has connection with an evacuation flange produced from polyamide at the end of the glass tube. Dried solid particles are shipped from the evacuation funnel. While infrared heat set at 85°C, between 800 and 400 μm sieve mesh sized 500 g powder has charged to the back funnel of the designed fluidized bed drying machine. At the end of 5 min, the evacuated powder is weighted and the moisture loss is calculated. At the end of the five repeated tests 9.8 wt.% moisture loss for beech powder and 5.3 wt.% moisture loss for pine wood powder in average are calculated [26].

In this work, 100 g dried wood powder is left 36 h at the room conditions (30% RH, 23°C) and moisturizing behavior is determined. At the end of the time period, beech wood powder moistened at 2.27 wt.% [5]. Graphical representation of the beech wood and pine wood powder weight rise by moistening against the time is given in **Figure 5**. Pine wood sawdust weight rise is determined as 1.25 wt.%.

Figure 5. Beech wood sawdust and pine wood sawdust weight rises according to time.

3. Manufacturing of WPC parts

Product form, process performance attributes, costs and the easiest way of manufacturing are the basic factors for the selection of appropriate process technology of the WPCs. In manufacturing operations, it is a must to understand the interrelationship between the materials, the processing technique and the final product design like being in WPCs production. The primarily concerns of the selection of process technology include, but are not limited to, shaping the product, well-distributing the wood fiber within the matrix, encapsulating the wood fiber and high adhesion of wood filler to the matrix resin. This interrelationship has a direct effect on the quality of the composite. The majority of the WPC applications consider recycling as well. This basic requirement has primarily driven by the developed processing technologies for WPC production. Extrusion, injection molding and compression molding processing technologies represent these areas of development [1].

On the other hand, many investigators have reported studies on the reaction of wood with a large number of alcohols to supply high process performances, subsequently. When a simple chemical pretreatment is used to soften the wood, defibrated wood structure can be achieved easily. The wood particles with the highest aspect ratio (is the ratio of fiber length to diameter) are produced with pretreated and undried wood chips, which also had the highest flexural and impact strength properties [33]. It is proved in a research work that to achieve consistent feeding into the extruder which enabled manufacturing of highly filled cellulose fiber–PP composites could be possible by pelletizing of the cellulose fibers [13].

Among the production techniques, extrusion is the process by which heat and shear forces are applied to a polymer within the barrel of an extruding machine. Extruders are generally characterized by L/D ratio. In an extrusion process, several operations are carried out similar to compounding process. These are feeding, heating, conveying-mixing, melting, devolatization and pumping processes. Counter-rotating twin screw extruders are widely being used either in parallel or conical type. For solid profiles with less demanding application requirements, single screw extruders are more commonly being used. Although WPCs are in general compatible with existing cost-effective processing machines, in practice, using these machines to produce WPCs cause some important processing problems. In the cases of using high percentage LC filler materials, the WPC certainly will represent different rheological characteristics during the process. Since screw elemental process machines are required special screw designs and certain chemical lubricating and bonding additives, users need to re-design the extruders specifically as compounder type. So that, wood-based reinforcement percentage of the WPC composites could be increased. For instance, at first, LC material should be preprocessed instead of using it directly in WPC production to overcome many reported processing difficulties which cause low productivity and poor product quality.

The primary processing difficulties are high die-head pressures, low moisture content requirement of LC material, high energy consumption and low mechanical and thermal efficiencies. High production costs and low productivity would probably be experienced as a result of these main difficulties. In this context, researchers focus on the investigation of wood-polymer composites compounding and production performances effective on production abilities and productivity. Compounding refers to an intimate mixing including dispersive and then distributive units in addition to the standard extrusion operations.

Several different configurations of twin screw extruders are suitable for achieving the desired final attributes and in executing the sequence of unit operations. General machine configurations are single-screw extruders, conical or parallel counter-rotating twin-screw extruders and planetary type screw extruders. Such machines used in the plastic industry cannot be used in the LC material–reinforced polymer composite production unless they are specifically re-designed to meet the requirements of WPC materials processing requirements. If the screw elemental process machines are going to be used especially for WPC manufacturing, high speed mixing and certain chemical lubricating and bonding are required. Sections of the screw may be re-designed to enhance compounding or blending, using lugging tools rather than a continuous screw flight. Additional hoppers should be assembled to feed the wood filler and other additives such as colorant or lubricant masterbatchs, antimicrobials or fire retardants. Port areas or venting channels are needed to allow steam and volatile gases to exit the system. They are generally assembled to the extruders used in WPC production to avoid blows or defects in the extruded part during processing. There are many other aspects to consider in extruder selection, starting from screw geometry to operating parameters which influence the flow, level of mixing and many other product properties such as feed rate, screw speed and barrel temperature [2, 5, 40].

It is an essential need to consider the quality of the products. In the extruder, the most important component is the screw, which is divided into three categories. The first is the feeder section where feeding takes place, the second is the transition section where melting happens and the third is the metering section where the product is pushed at a uniform rate and pressure through the die [22]. The extruder should be capable of consistently deliver the thermally homogeneous material to the die at high pressures. Die design is critical to ensure uniform melt delivery across the die to minimize cross-channel velocity gradients and resulting physical property gradient effects because of the complex content and high viscosity of the composites. Die design may be quite complex especially for hollow sectioned profiles of many kinds of geometries and various materials having different rheology and shrinkage behaviors. The nature of matrix resins affects total shrinkage behavior of the composite. Die entrance and exit angel designs are the key factors affecting the mechanical properties of the composites [1].

Injection molding process is similar to metal die casting in most respects. It is used to produce many kinds of products from toys to automobile parts. Injection molding is nearly for one third of polymeric material processing types as being a versatile process. The aim of conventional injection molding is the mass production of parts with complex geometries at low costs and high production rates. However, the range of materials is limited with thermoplastic or thermoset polymers. In the case of using natural fibers with high aspect ratios in automotive applications, reaction injection molding, structural reaction injection molding and reinforced reaction injection molding processes are very common. Modern injection molding machines use the reciprocating screw for plasticization, the same as for extrusion, but the screw is stopped and then being used as a plunger. Plunging injects the hot polymer into the mold cavity [22].

To sum up, most of the composites used in constructive applications are extruded to a profile of uniform solid or hollow cross section at any length, whereas products having more complex shapes such as being in the automotive industry or other consumer products are injection or compression molded [41]. Products such as decking, railings and window or door profiles readily exude themselves to extrusion through a two-dimensional die. Injection-molded applications such as consumer household goods and furniture parts are gaining importance due to the high consumption rates of these kinds of products. Thermoforming or compression molding method is especially preferable for automotive industry [4]. Besides, microcellular injection molding process is a novel method using supercritical gases mixed with polymer melt in the barrel and then a single-phase polymer gas solution is created. This process can find wide application areas in WPC production especially for the automotive industry.

4. An experimental work on WPC production

Commercially available virgin low density polyethylene (LDPE) was used in the conducted experimental works. The basic properties of the LDPE are; 20 g/10 min. MFI at 190°C/2.16 kg, 0.923 g/cm^3 density and 10.5 MPa tensile strength. The used additives were 1 wt.% slip and anti-blocking agent and 1 wt.% as coupling agent. Similarly, commercially available beech wood powder was provided from one of the lumber market. The filtered and sieved beech wood powder was operated in the pulping bath, consisting of 10% Na_2SO_3 (wt.%), for 30 min at 95°C [6, 28]. After grinding, the processed filler wood material is mixed with the virgin polymer. The compound ratio was 50 wt.% of pulped wood reinforcement and 48 wt.% LDPE.

After mechanical mixing, the mixture was transferred to the constructed compounding extruder. A compounder type extruder was designed for the production of WPC considering polymer feeding sections, heating and distributing parts with an adapted melt pump system for the compression. The extruder is a single-screw compounding machine. The barrel has a diameter of 30 mm, a length of 800 mm and consists of 6 zones with separate temperature control units. The stable thermal conditions practiced for WPC production are listed in **Table 2**.

Heat zones	1	2	3	4	5	6
T (°C)	120	130	140	140	130	120

Table 2. Thermal conditions practiced for WPC production.

As the processing parameters, the extruder screw speed was set to 110 rpm at 20 Hz frequency and the machine torque was changed 0.37 to 0.57 Nm during the processes. Schematic illustration of the designed compounding extruder is given in **Figure 6**.

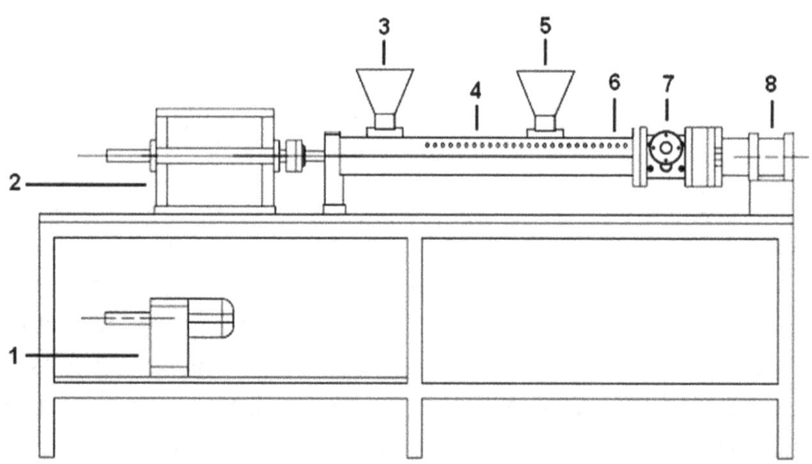

No	Description
1	Electrical Motor
2	Reduction Gear
3	First Feed Hopper
4	Single Screw
5	Second Feed Hopper
6	Barrel
7	Spin Pump
8	Extrusion Die

Figure 6. Schematic illustration of the compounding extruder.

The maximum operating temperature was 140°C at all trials. Although polyolefin is a viscose material, the transferred heat to the LC material affects the viscosity of the WPC in a negative way. The required low viscosity of the materials enables processing at low temperatures. So that, extruding requires low pressure and could enhance high productivity by the reduction in the cycle time. Due to the low temperature and low pressure requirements, energy consumption and thermal degradation of natural fillers can be reduced as well. As observed in the experiments, giving viscose friction energy to the LC material throughout the extrusion process is not only wasting the energy but also causing chemical deformation of the material. That is why researchers are oriented to modify manufacturing techniques for LC material filled polymer composite production instead of using conventional machines.

In this experimental work, pulped and dried wood powder is delivered from the first feed hopper of the compounding extruder to the fluidized bed by the screw to mix with polyolefin. After that, polyolefin and LC are mixed homogeneously by the screw at lower squeezing and higher mixing rates. Finally, the mix is pressurized by a gear pump to manufacture the final profile at the die. The die gravure is designed for circular bars. After cutting this raw material into small pieces, the final product is manufactured in a single screw extruder. The pictures of circular bars as raw WPC material and trim boards as final WPC products can be seen in **Figure 7**, respectively.

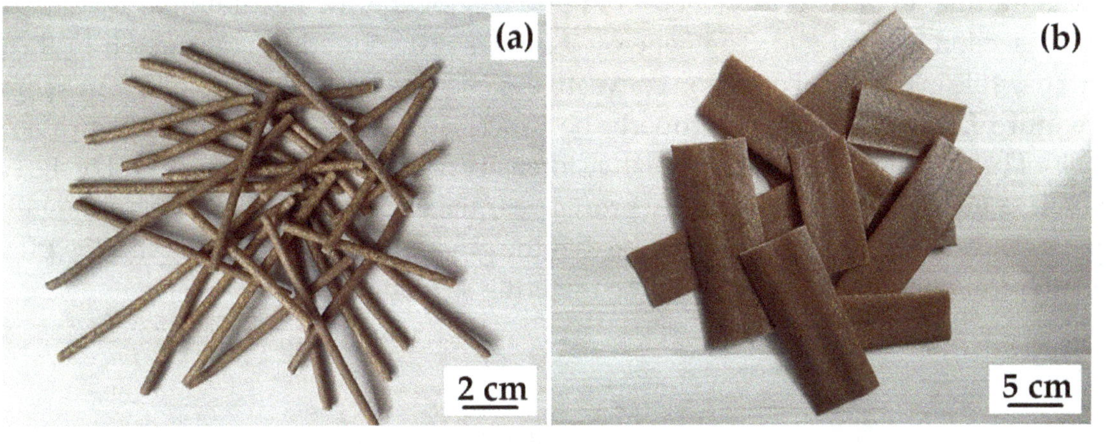

Figure 7. Pictures of (a) circular bars as raw WPC material and (b) trim boards as final WPC products.

In order to find out the mechanical properties of manufactured parts, tensile tests were conducted in accordance with the ISO 527-2 standard on a Schimadzu tensile tester, Japan, AG-X type with a load cell of 100 kN and a universal extensometer [42]. The flexural tests were carried out according to the ISO 178 standard [43]. The repeated mechanical test results of the manufactured composites are given in **Table 3**. It is observed that mechanical qualifications of the produced wood-polymer composites have been enhanced regarding to the virgin PE. With the help of pelletization process and machine improvements, high productivity manufacturing was succeeded.

Tests	Flexural strength (MPa)	Elongation at flexural break (%)	Tensile strength (MPa)	Elongation at tensile break (%)
1st	48.80	3.11	16.75	0.62
2nd	54.20	1.63	16.95	0.90
3rd	54.50	2.90	18.05	1.80
4th	43.10	2.21	18.25	1.85
5th	50.10	2.82	17.70	1.60
Mean	**49.94**	**2.53**	**17.54**	**1.35**
St. deviation	**4.65**	**0.61**	**0.66**	**0.56**

Table 3. Mechanical test results of the manufactured composites.

Since screw design is very important for WPC production, several simulations should be performed on screws according to the expected attitudes from them. In this experiment, the authors performed some simulations on the screw's longitudinal plane considering pressure and velocity flows by ANSYS CFD 13.0 software program. The simulation method is based on computational fluid dynamics (CFD) refers to the use of the numerical techniques to solve fluid dynamical problems. Numerical techniques consist of a very broad range of methods including, but not limited to, finite difference methods (FDM), finite element methods (FEM), finite volume methods (FVM) and boundary element methods (BEM). The pressure contours on a longitudinal plane is represented in **Figure 8**, showing the maximum at the end of the metering section as expected. The generated heat is transported by convection in the axial direction; while in the radial direction convective exchanges are limited, resulting in a complex temperature field which cannot properly be described using a single normal profile. The analysis of velocity field shows that a relative intensity mixing is present at the end of the screw, as shown in **Figure 9**. The flow velocity profile near the barrel wall is high, decreasing to zero at the screw root. The figures also indicate that the gradients for the velocity and the pressure show a typical polymer flow profile in the metering section.

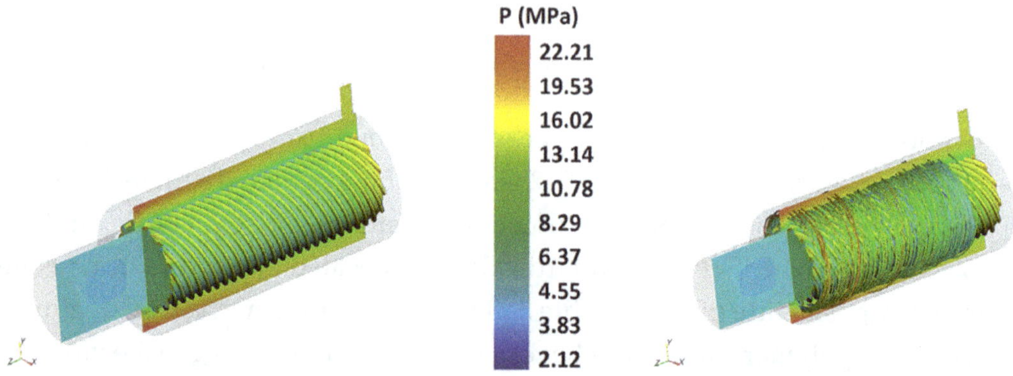

Figure 8. Pressure contours on a longitudinal plane.

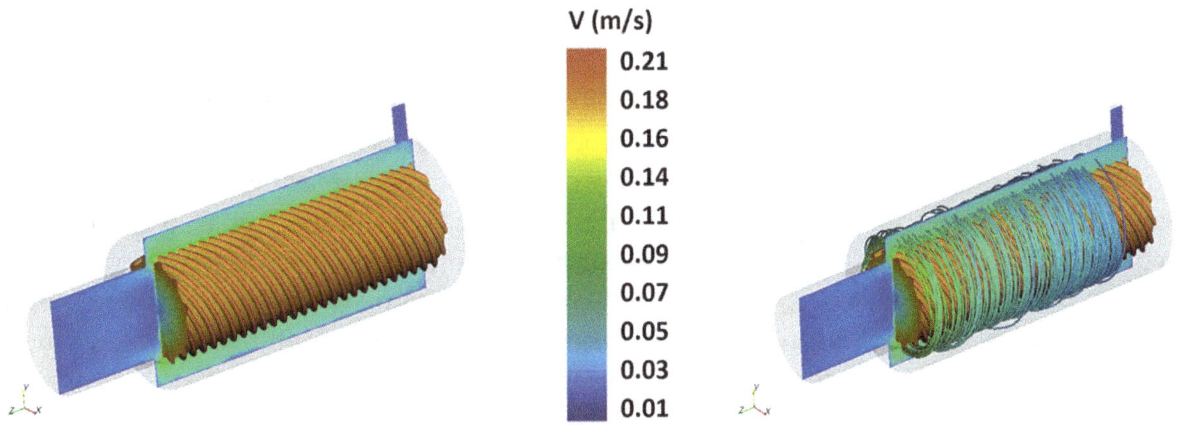

Figure 9. Velocity contours on a longitudinal plane.

5. Conclusions and future works

LC materials used in WPC productions can be derived from a variety of natural sources. Geographical resources are the drivers of filler material type and choice. It is clear that the use of LC fibers will continue to rise. However, the type of fiber in a geographic location is dictated by the factors including, but not limited to, availability, price, yield and required mechanical properties.

Wood powder is commercially processed from industrial waste materials such as planer shavings, chips and sawdust. During WPC production, the addition of wood powder into the plastic matrix significantly reduces the tensile and flexural strengths of the composite due to the poor interfacial adhesion between wood powder and the matrix in case not to using of coupling agents. The addition of coupling agents has many positive effects on the strength of WPC parts supplying an enhanced interfacial adhesion between wood powder and plastic matrix.

Nowadays, there are various methods to improve the most important usage performances of WPC parts. Besides stiffness, wood fiber reinforcement improves creep resistance, bending and tension abilities of WPC parts. WPC is less capable of necking than virgin polymer, so strain at break is lower and the Young's modulus is higher due to the stiffening effect of the wood particles; however, improvement in tensile strength is not guaranteed, whereas bending strength is generally higher. To increase the strength and modulus values of tensile and flexural tests, the use of compatibilizer is inevitable. The increasing usage of coupling agents would aid further cost decreases in WPCs. New LC sources include paper mill sludge and biorefinery residues. Because wood and LC fibers can lead to superior WPC properties by acting more as reinforcement than as filler, the present trend is to move toward the use of wood and other LC fibers in additional strength requirements.

All sources of variation in manufacturing operations are accepted as crucial. In order to optimize the complete manufacturing line and the process for WPC products, it is essential to operate each step of the manufacturing and all process parameters by well-characterized design of experiment techniques and methods. Subsequently, with the use of in wide variety optimization techniques, the best combinations of all operations can minimize downtime and costs while maximizing productivity and composite performance characteristics even considering the environment. Moreover, production technologies increasingly shift in favor of composites especially when life cycle costs are inevitably considered. Today, techniques are needed to develop for many applications. Improvements and changes in manufacturing methods will help to make preparation for the next generation WPCs. New processes under development are coextrusion and foaming during extrusion process. Inline coating technologies are one of the primary interests of the researchers nowadays.

Long life time resistance of the WPCs at many applications continues to be an issue of the researchers. Specifically, moisture uptake targeting is being considered in developing technologies. Commercially available chemicals or additives that absorb water, thereby limiting degradation of the WPC, are on the rise, especially for outdoor applications. Moreover, novel composites, nanocomposites, including the use of nanofibers derived from wood and other LC materials are taking great attention recently. Finally, concerning the environmental awareness of the users, researchers are working on completely green composite materials requiring the development of polymers that are totally biodegradable and environment friendly. WPCs are not only facing present requirements of consumers but also will high probably meet their further concerns by conducting developments in the field.

Author details

Anil Akdogan* and Ali Serdar Vanli

*Address all correspondence to: nomak@yildiz.edu.tr

Mechanical Engineering Department, Yildiz Technical University, İstanbul, Turkey

References

[1] Shankar G. Thermoplastic wood fiber composites. In: Mohanty A K, Misra M, Drzal L T, editors. Natural fibers, biopolymers and biocomposites. 1st ed. Boca Raton: CRC; 2005. pp. 348–386.

[2] Spear M J, Eder A, Carus M. Wood polymer composites. In: Ansell M P, editor. Wood composites. 1st ed. Cambridge: Woodhead; 2015. pp. 195–249. doi:10.1016/B978-1-78242-454-3.09994-7

[3] Najafi S K. Use of recycled plastics in wood plastic composites—a review. Waste Management. 2013;33(9):1898–1905. doi:10.1016/j.wasman.2013.05.017

[4] Matuana L M, Stark N M. The use of wood fibers as reinforcements in composites. In: Faruk O, Sain M, editors. Biofiber reinforcements in composite materials. 1st ed. Cambridge: Woodhead; 2015. pp. 648–688. doi:10.1533/9781782421276.5.648

[5] Akdogan A, Vanli A S. Material characterization of different-dimensioned wood particle-reinforced polymer composites. Journal of Thermoplastic Composite Materials. 2013;26(9):1237–1248. doi:10.1177/0892705713484743

[6] Sykacek E, Hrabalova M, Frech H, Mundigler N. Extrusion of five biopolymers reinforced with increasing wood flour concentration on a production machine, injection moulding and mechanical performance. Composites Part A: Applied Science and Manufacturing. 2009;40(8):1272–1282. doi:10.1016/j.compositesa.2009.05.023

[7] Sarul I T, Akdogan A, Koyun A. Alternative production methods for lignocellulosic composite materials. Journal of Thermoplastic Composite Materials. 2010;23(3):375–384. doi:10.1177/0892705709345954

[8] Hietala M, Samuelsson E, Niinimaki J, Oksman K. The effect of pre-softened wood chips on wood fibre aspect ratio and mechanical properties of wood-polymer composites. Composites Part A: Applied Science and Manufacturing. 2011;42(12):2110–2116. doi: 10.1016/j.compositesa.2011.09.021

[9] Islam M S, Hamdan S, Hasan M, Ahmeda A S, Rahman M R. Effect of coupling reactions on the mechanical and biological properties of tropical wood polymer composites (WPC). International Biodeterioration & Biodegradation, 2012;72:108–113. doi:10.1016/j.ibiod.2012.05.019

[10] Bengtsson M, Baillif M L, Oksman K. Extrusion and Mechanical properties of highly filled cellulose fiber-PE composites. Composites Part A: Applied Science and Manufacturing. 2007;38:1922–1931. doi:10.1016/j.compositesa.2007.03.004

[11] Morreale M, Scaffaro R, Maio A, La Mantia F P. Effect of adding wood flour to the physical properties of a biodegradable polymer. Composites Part A: Applied Science and Manufacturing. 2008;39(3):503–513. doi:10.1016/j.compositesa.2007.12.002

[12] Reedy M. New chemical foaming agents expand wood/plastic composite market. Plastics, Additives & Compounding. 2002;4(5):24–26. doi:10.1016/S1464-391X(02)80114-3

[13] Felix J S, Domeno C, Nerin C. Characterization of wood plastic composites made from landfill-derived plastic and sawdust: volatile compounds and olfactometric analysis. Waste Management. 2013;33:645–655. doi:10.1016/j.wasman.2012.11.005

[14] Jam N J, Behravesh A H. Flow behavior of HDPE-fine wood particles composites. Journal of Thermoplastic Composite Materials. 2007;20:439–451. doi:10.1177/0892705707082324.

[15] Kücük M, Demirbas A. Kinetic study on hydrolysis of biomass (ailanthus altissima chips) by using alkaline-glycerol solution. Energy Conservation and Management. 1999;40(13):1397–1403. doi:10.1016/S0196-8904(99)00031-X

[16] Nygard P, Tanem B S, Karlsen T, Brachet P, Leinsvang B. Extrusion-based wood fiber-pp composites: wood powder and pelletized wood fibers—a comparative study. Composites Science and Technology. 2008;68(15–16):3418–3424. doi:10.1016/j.compscitech.2008.09.029

[17] Gosselin R, Rodrigue D, Riedl B. Injection molding of postconsumer wood-plastic composites I: morphology. Journal of Thermoplastic Composite Materials. 2006;19:639–657. doi:10.1177/0892705706067484

[18] Gosselin R, Rodrigue D, Riedl B. Injection molding of postconsumer wood-plastic composites II: mechnacial properties. Journal Of Thermoplastic Composite Materials. 2006;19:659–669. doi:10.1177/0892705706067486

[19] Schirp A, Mannheim M, Plinke B, Influence of refiner fibre quality and fibre modification treatments on properties of injection-moulded beech wood–plastic composites. Composites: Part A. 2014;61:245–257. doi:10.1016/j.compositesa.2014.03.003.

[20] Akhtar M, Scott G, Swaney R, Shipley D. Biomechanical pulping: a mill-scale evaluation. Resource, Conservation and Recycling. 2000:28:241–252. doi:10.1016/S0921-3449(99)00048-8

[21] Danyadi L, Janecska T, Szabo Z, Nagy G, Moczo J, Pukanszky B. Wood flour filled PP composites: compatibilization and adhesion. Composites Science and Technology. 2007;67(13):2838–2846. doi:10.1016/j.compscitech.2007.01.024

[22] Bledzki A K, Letman M, Viksne A, Rence L. A comparison of compounding processes and wood type for wood fibre-PP composites. Composites Part A: Applied Science & Manufacturing. 2005;36(6):789–797. doi:10.1016/j.compositesa.2004.10.029

[23] Colling D A, Vasilos T. Industrial materials, volume 2: polymers, ceramics and composites. 1st ed. Columbus: Pearson; 1995. p. 256.

[24] Seethamraju K V, Beaverson N J, Heikkila K E. Advanced compatible polymer wood fiber composite. Andersen Corporation. US Patent Number 5981067-A; 9 November 1999.

[25] Carpenter M J, Sweetman G W. Tool for coldforming. Sandvik Intellectual Property Ab. US Patent Number 5948523-A; 7 September 1999.

[26] Vanli A S, Akdogan A, Koyun A. Fluidized bed drying system for continuous processing of lignocellulosic polymer composite materials. In: Proceedings of International Journal of Arts and Sciences Conference (IJAS '09); 9–13 November 2009; Gottenheim. New Britain: IJAS. 2009;3(1):43–49.

[27] Law K N, Kokta B V, Mao C B. Fibre morphology and soda - sulphite pulping of switchgrass. Bioresource Technology. 2001;77(1):1–7. doi:10.1016/S0960-8524(00)00140-1

[28] Demirbas A. Aqueous glycerol delignification of wood chips and ground wood. Biosource Technology. 1998;63(2):179–185. doi:10.1016/S0960-8524(97)00063-1

[29] Baroulaki I, Karakasi O, Pappa G, Tarantili P A, Economides D, Magoulas K. Preparation and study of plastic compounds containing polyolefins and post used newspaper

fibers. Composites Part A: Applied Science and Manufacturing 2006;37(10):1613–1625. doi:10.1016/j.compositesa.2005.10.012

[30] Van Krevelen D W, Te Nijenhuis K. Properties of polymers: their correlation with chemical structure; their numerical estimation and prediction from additive group contributions. 4th ed. Amsterdam: Elsevier; 2009. p. 1032. doi:10.1016/B978-0-08-054819-7.00043-1

[31] Gupta B S, Reiniate I, Laborie M P G. Surface properties and adhesion of wood fiber reinforced thermoplastic composites. Colloids and Surfaces A: Physicochemical and Engineering Asppects. 2007;302(1–3):388–395. doi:10.1016/j.colsurfa.2007.03.002

[32] Markarian J. Additive development aid growth in wood plastic composites. Plastics, Addities & Compounding. 2002;4(11):18–21. doi:10.1016/S1464-391X(02)80174-X

[33] Markarian J. Wood-plastic composites: current trends in materials and processing. Plastics, Additives & Compounding. 2005;7(5):20–26. doi:10.1016/S1464-391X(05)70453-0

[34] Keener T J, Stuart R K, Brown T K. Maleated coupling agents for natural fiber composites. Composites Part A: Applied Science and Manufacturing. 2004;35(3):357–362. doi: 10.1016/j.compositesa.2003.09.014

[35] Nourbakhsh A, Karegarfard A, Ashori A, Nourbakhsh A. Effects of particle size and coupling agent concentration on mechanical properties of particulate-filled polymer composites. Journal of Thermoplastic Composite Materials. 2010;23(2):169–174. doi: 10.1177/0892705709340962

[36] Leu S Y, Yang T H, Lo S F, Yang T H. Optimized material composition to improve the physical and mechanical properties of extruded wood–plastic composites (WPCs). Construction and Building Materials. 2012;29:120–127. doi:10.1016/j.conbuildmat. 2011.09.013

[37] Migneaulta S, Koubaab A, Perrec P, Riedl B. Effects of wood fiber surface chemistry on strength of wood-plastic composites. Applied Surface Science. 2015;343:11–18. doi: 10.1016/j.apsusc.2015.03.010

[38] Markarian J. Process modifiers improve output and cost competitiveness. Plastics, Additives & Compounding. 2006;8(6):20–23. doi:10.1016/S1464-391X(06)70660-2

[39] Rizvi G, Matuana L M, Park C B. Foaming of PS/wood-fiber composites in extrusion using moisture as a blowing agent. Polymer Engineering & Science. 2000;40(10):2124–2132. doi:10.1002/pen.11345

[40] Matuana L M, Mengeloglu F. Manufacture of rigid PVC/wood-flour composite foams using moisture contained in wood as foaming agent. Journal of Vinyl & Additive Technology. 2002;8(4):264–270. doi:10.1002/vnl.10373

[41] Yeh S K, Gupta R K. Improved wood-plastic composites through better processing. Composites Part A: Applied Science and Manufacturing. 2008;39(11):1694–1699. doi: 10.1016/j.compositesa.2008.07.013

[42] ISO 527-2: 2012. Plastics—determination of tensile properties—part 2: test conditions for moulding and extrusion plastics, 2nd ed. International Standard Published; 2012; p. 11.

[43] ISO 178: 2010. Plastics—determination of flexural properties, 5th ed. International Standard Published; 2011; p. 19.

5

Wood Thermal Properties

Richard Hrčka and Marián Babiak

Abstract

Wood thermal properties specify the answers to the questions related to heat transfer. The values of specific heat, thermal conductivity, and thermal diffusivity were simultaneously determined with quasistationary method. Wood is distinguished as a natural material for accumulating the energy by heat transfer, as isolating material, with the ability to slowly equilibrate its different temperatures. The measured thermal properties value of beech and fir wood samples support those conclusions. Known dependences of wood thermal properties on anatomical direction, density at given moisture content, temperature are modelled and incorporated into heat conduction equation to provide base for next evaluation of measured data. The heat conduction solutions, based on known wood thermal properties, are used in similar problems. It is shown that thermal properties influence the surface equilibrium temperature between skin and wooden sample and the solution of heat conduction equation describes the flux passing through the log as an element of log-cabin house. Also thermal diffusivity is a component of equation that determines the position of observed point of wood during conduction. The results served as a base point for planning the experiments, for designing the processes of heat transfer, for designing the furniture and wooden houses, for designing the machines and equipment in woodworking industry and others.

Keywords: wood, specific heat, thermal conductivity, thermal diffusivity, coefficient of thermal expansion

1. Introduction

The desired change of wood temperature is involved in wood processing such as drying, forming, gluing, finishing and others. Also, suitable temperature is a part of comfortable environment in wooden houses. The heat transfer is one of the processes how to change the temperature of wood. Wood is surrounded by boundary from its surrounding. Heat transfer occurs spontaneously through the boundary solely due to the non-zero difference in temperature of wood and its surrounding. The processes by which the heat is transferred are classified into three categories: conduction, convection and radiation. The mechanism of conduction is dominant process of heat transfer through wood in many previously mentioned situations of wood processing. The convection and radiation are also included in description of heat transfer through wood, mainly in the form of boundary conditions. The basic condition for convection is moving medium. Relative motion of wood and surrounding occurs in convective boundary condition. Also, permeable wood transfers simultaneously work and heat (outside pressure and temperature as potential). Such mechanism produces deviation from results of conduction.

We are solving three basic questions related to heat transfer by wood:

1. How much heat is needed for changing the wood temperature by 1 K?
2. How much heat is flowing through the wall?
3. How fast two different temperatures in wood equilibrate?

The aim of the chapter is to answer to these questions. The first question is related to specific heat capacity, the second question is related to thermal conductivity and the third one to thermal diffusivity. The differences in wood temperature at constant pressure causes wood dimensional changes. The question about wood dimension after wood temperature change is related to thermal expansion coefficient. However, this question is often omitted because small value of thermal expansion coefficient in comparison with coefficient of swelling or shrinking. All the mentioned quantities are measured for wood. The quantities definitions are independent from different measuring methods, but often assign quantities values with appropriate units do not keep such clearness. Therefore, the next part of chapter is devoted to measuring method of wood thermal properties.

2. Research method and results

Heat, internal energy, entropy are hardly measured quantities [1]. But temperature, length, mass and time are easy measureable. Therefore, heat is computed from easier measured quantities, for example, temperature difference and others. Let wood be a system of conserving the enthalpy H:

$$H = U + pV \tag{1}$$

where U is internal energy, p is pressure, V is volume of wood. The next equation expresses continuity for enthalpy:

$$\frac{\partial H}{\partial t} + \frac{\partial q}{\partial x} = 0 \tag{2}$$

where q is heat flux, t is time and x is dimensional coordinate. After applying the first thermodynamic principle to equation of enthalpy change, it follows:

$$\Delta \mathrm{H} = \mathrm{Q\text{–}W'} + \Delta(\mathrm{pV}) = \mathrm{mc_p}\Delta\mathrm{T} + \mathrm{V}\Delta\mathrm{p} = (\rho c_{\mathrm{p}}\Delta\mathrm{T} + \Delta\mathrm{p})\mathrm{V} \tag{3}$$

where Q is heat, W′ is work carried out by wood on surrounding, m is wood mass, c_p is wood mass specific heat capacity at constant pressure, ΔT is temperature difference in wood. Specific heat is the amount of heat transferred to unit mass of wood to rise its temperature by 1 K. Specific heat is property by which we are able to distinguish wood as a good material for accumulate the energy by heat transfer. The flux which changes the enthalpy is divided into two parts: flux which is related to change of temperature q_U, and flux which is related to change of pressure q_W inside arbitrary wood infinitesimal volume dV:

$$\left(\rho c_P \frac{\partial T}{\partial t} + \frac{\partial q_U}{\partial x}\right) dV = 0 \tag{4}$$

$$\left(\frac{\partial p}{\partial t} + \frac{\partial q_W}{\partial x}\right) dV = 0 \tag{5}$$

where ϱ is wood density. The sum of both previous equations gives zero sources of enthalpy in wood volume. Both equations are valid for arbitrary volume and, moreover, there is need to determine the flux of internal energy as temperature function.

It follows from observation, that temperature difference in space spontaneously produces the heat rate from higher to smaller values of temperature. The heat rate Q/t is proportional to temperature difference ΔT, to the area of heat transfer S and inversely proportional to the distance in space d. These findings are summarized by Fourier law:

$$\frac{Q}{St} = \lambda \frac{\Delta T}{d} \tag{6}$$

where λ is thermal conductivity and represents the wood property to conduct heat. Thermal conductivity is property by which we are able to distinguish wood as heat insulator or pure heat conductor. The differential form of Fourier law:

$$\vec{q} = -\bar{\bar{\lambda}} grad(T) \tag{7}$$

relates the heat flux q and the gradient of temperature. In general, vector of heat flux and temperature gradient are not at one line, therefore, thermal conductivity is the second order tensor. Its eigenvalues are positive numbers in $Wm^{-1} K^{-1}$. The minus sign in Eq. (7) expresses the increase of the wood temperature by the heat flux directed to its volume, which surface is oriented outside the volume. We have always made sure with measurement that thermal conductivity is symmetric tensor and; moreover, for wood, it is possible to arrange it to diagonal form according to suitable transformation. The form of Fourier law for radial slab is:

$$\begin{pmatrix} q_z \\ q_x \\ q_y \end{pmatrix} = -\begin{pmatrix} \lambda_L & 0 & 0 \\ 0 & \lambda_R & 0 \\ 0 & 0 & \lambda_T \end{pmatrix} \begin{pmatrix} \frac{\partial T}{\partial z} \\ \frac{\partial T}{\partial x} \\ \frac{\partial T}{\partial y} \end{pmatrix} \tag{8}$$

where wood anatomical directions (L—longitudinal, R—radial, T—tangential) coincide with orientation of Cartesian coordinate axis (z, x, y). Radial board is linear orthotropic material.

The form of Fourier law for tangential slab is formulated:

$$\begin{pmatrix} q_z \\ q_x \\ q_y \end{pmatrix} = -\begin{pmatrix} \lambda_L & 0 & 0 \\ 0 & \lambda_R \dfrac{x^2}{x^2+y^2} + \lambda_T \dfrac{y^2}{x^2+y^2} & (\lambda_R - \lambda_T)\dfrac{xy}{x^2+y^2} \\ 0 & (\lambda_R - \lambda_T)\dfrac{xy}{x^2+y^2} & \lambda_T \dfrac{x^2}{x^2+y^2} + \lambda_R \dfrac{y^2}{x^2+y^2} \end{pmatrix} \begin{pmatrix} \dfrac{\partial T}{\partial z} \\ \dfrac{\partial T}{\partial x} \\ \dfrac{\partial T}{\partial y} \end{pmatrix} \tag{9}$$

Fourier law for log is suitable to express in cylindrical coordinates $(z, r, \phi) = (L, R, T)$:

$$\begin{pmatrix} q_z \\ q_r \\ q_\varphi \end{pmatrix} = -\begin{pmatrix} \lambda_z & 0 & 0 \\ 0 & \lambda_r & 0 \\ 0 & 0 & \lambda_\varphi \end{pmatrix} \begin{pmatrix} \dfrac{\partial T}{\partial z} \\ \dfrac{\partial T}{\partial r} \\ \dfrac{\partial T}{r\partial \varphi} \end{pmatrix} \tag{10}$$

The log is cylindrical orthotropic material. Fourier law (7) does not contain time of the process explicitly. The coupling of Fourier's law and Eq. (4) results in heat conduction equation:

$$\mathrm{div}(\overline{\overline{\lambda}}\,\mathrm{grad}(T)) + s = c_p \rho \frac{\partial T}{\partial t} \tag{11}$$

s denotes the rate of energy release or its consumption in volume unit of internal sources or sinks. As wood can be distinguished as cylindrical orthotropic material [2–4] or as its special case linear orthotropic material, heat conduction equation has the form with constant eigenvalues of thermal diffusivity α:

$$\frac{\alpha_R}{r}\frac{\partial}{\partial r}\left(r\frac{\partial T}{\partial r}\right) + \frac{\alpha_T}{r^2}\frac{\partial^2 T}{\partial \varphi^2} + \alpha_L \frac{\partial^2 T}{\partial z^2} = \frac{\partial T}{\partial t} \tag{12}$$

in cylindrical coordinate system as wood has the form of log or:

$$\alpha_R \frac{\partial^2 T}{\partial x^2} + \alpha_T \frac{\partial^2 T}{\partial y^2} + \alpha_L \frac{\partial^2 T}{\partial z^2} = \frac{\partial T}{\partial t} \tag{13}$$

in Cartesian system as wood has the form of radial slab. Thermal diffusivity is the ratio of thermal conductivity and product of mass specific heat capacity and density at given moisture content. Thermal diffusivity is property by which we are able to distinguish that wood equilibrates its different temperatures slowly. Also, it describes the slowest temperature change in wooden body. The solution of heat conduction equation provides the temperature field in wood (direct problem) and the base of measurement method of thermal properties (inverse problem). Advantage of solutions is possibility to describe the similar examples of heat transfer by conduction. The various techniques (analytical or numerical) are employed to solve heat conduction equation. Both of them have the same feature of using initial and boundary conditions to compute particular solution. The initial and boundary conditions should match the situation as best as possible. The initial condition describes the temperature field in wood at the beginning of the situation. There is possibility to recognize three kinds of boundary conditions or their combination [5, 6]:

the 1st (I.) kind boundary condition describes the surface temperature as a function of time (Dirichlet condition),

the 2nd (II.) kind boundary condition describes the heat flux at the surface as a function of time (Newman condition),

the 3rd (III.) kind boundary condition describes the heat flux at the surface as a function of surface temperature (Robin condition).

The constant temperature at the surface represents the 1st kind boundary condition. It approaches the process of wood pressing, when press hot metal tables as heat reservoir touch the wood surfaces [7]. Then the variable temperature at the surface can be modelled according the Duhamel theorem. If wood is heated by external radiant source, linear increase surface temperature in time of heating will describe the situation [8].

The constant heat flux at the surface represents the 2nd kind boundary condition. It approaches the wood heating, when metal table touching the wood surface is heated by electric current [9]. Also zero heat flux represents the adiabatic process or symmetry in temperature field.

The constant proportionality between heat flux at the surface and the surface temperature represents the 3rd kind boundary condition. It describes the surface phenomena during heat transfer when fluid touches the wood surface. Also, heat transfer between low temperature radiant sources and wood specimen fulfil this condition [5]. Then heat transfer coefficient is proportional to emissivity of wood.

The coupling of the first and second boundary conditions enables to exclude the time from boundary condition. Such situation occurs at the wood surface touching the solid [10].

The application of similar planar heat source as used by [9] in their unsteady state method and apparatus arrangement of method as used by [11] continues to method used in the Department of Wood Science at Technical University in Zvolen. The sample arrangement of the method, named as quasistationary method [2], is depicted at **Figure 1**.

The eight specimens' arrangement is symmetric which fulfil the following boundary conditions:

$$\left.\frac{\partial T}{\partial x}\right|_{x=d_R} = \frac{q}{\lambda} \tag{14}$$

$$\left.\frac{\partial T}{\partial x}\right|_{x=0} = 0 \tag{15}$$

$$T(x,0) - T_0 = 0 \tag{16}$$

where d_R is thickness of one specimen. Eq. (14) represents the constant flux at the surface $x = d_R$. The very thin (0.01 mm) NiCr foil is heated by direct electric current. It is produced

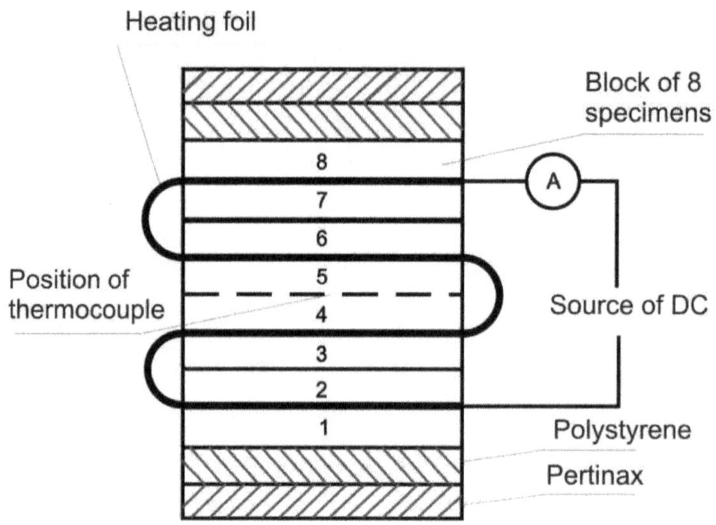

Figure 1. The diagram of quasistationary method.

by stable laboratory DC source. It is assumed the heat is symmetrically distributed to adjacent specimens, therefore, heat flux q is computed:

$$q = \frac{1}{2}\frac{RI^2}{S} \tag{17}$$

where R is resistance, I is direct current, S is rectangular area of one foil surface touching the specimen. Eq. (15) describes zero flux at the centre of the block of the eight specimens. The initial condition (16) prescribes constant temperature T_0 throughout the specimen at the beginning of experiment. Then, the solution of heat conduction equation in one dimension has the form:

$$T(x,t){-}T_0 = \frac{qd_R}{\lambda}\left[\frac{\alpha t}{d_R^2} - \frac{d_R^2{-}3x^2}{6d_R^2} + \sum_{n=1}^{\infty}(-1)^{n+1}\frac{2}{(n\pi)^2}\cos\left(n\pi\frac{x}{d_R}\right)e^{-(n\pi)^2\frac{\alpha t}{d_R^2}}\right] \tag{18}$$

The sum in Eq. (18) can be significantly active only at the beginning of the experiment. It is possible to neglect it for sufficient long time. And finally, the linear increase of temperature in the middle of the 8 block of specimens is:

$$T(0,t){-}T_0 = \frac{qd_R}{\lambda}\left(\frac{\alpha t}{d_R^2} - \frac{1}{6}\right) = At + B = \frac{q}{\rho d_R}\frac{1}{c}t - \frac{qd_R}{6}\frac{1}{\lambda} \tag{19}$$

If A is the slope and B is the intercept of such linear increase of temperature in time, the formulas are valid for thermal properties:

$$c = \frac{q}{Ad_R\rho} \tag{20}$$

$$\lambda = -\frac{qd_R}{6B} \tag{21}$$

$$\alpha = -\frac{d_R^2 A}{6B} \tag{22}$$

The method is named as quasistationary method because characteristic linear temperature increase in time is present. Thermal diffusivity can be determined, even the flux from source is not known. Moreover, if density is known, the specific heat is determined. The characteristic temperature increase in time for whole duration of experiment is in **Figure 2**.

The one dimensional adiabatic model (18) is proper for thin radial boards, thin specimen with the thickness in longitudinal direction or very distant tangential boards from the pith. Wood is treated as linear orthotropic material. The solution:

$$T(r,\varphi,t)-T_0 = \frac{qr\varphi_{max}}{\lambda_T}\left[\frac{\alpha_T t}{(r\varphi_{max})^2} - \frac{{\varphi_{max}}^2 - 3\varphi^2}{6{\varphi_{max}}^2} + \sum_{n=1}^{\infty}(-1)^{n+1}\frac{2}{(n\pi)^2}\cos\left(n\pi\frac{\varphi}{\varphi_{max}}\right)e^{-(n\pi)^2\frac{\alpha_T t}{(r\varphi_{max})^2}}\right] \tag{23}$$

where $r \neq 0$ m, is suitable for the block of 8 wooden wedges and wood is treated as cylindrical orthotropic material. The angle between the marginal radial surfaces of one wedge is φ_{max}. Then computing of thermal properties from temperature increase in time is the similar to previous formulas (20)–(22).

Models (18) and (23) are adiabatic, there is no lateral transfer of heat to surrounding there and, therefore, the models are one-dimensional. The three dimensional model enables to simultaneously predict all material thermal diffusivity eigenvalues in principal anatomical directions together with its specific heat. Such model should describe also lateral heat transfer to

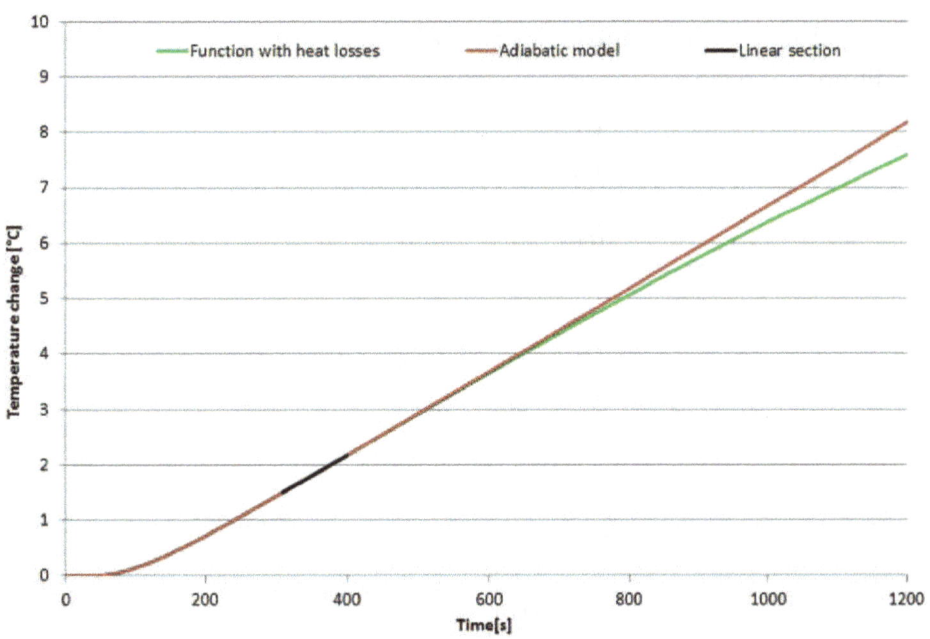

Figure 2. The adiabatic solution (18) together with the linear part t ε (300; 400)s typical for quasistationary method (α= 1.5×10^{-7} m^2 s^{-1}, c = 1.8 kJ kg^{-1} K^{-1}, ρ = 416 kg m^{-3}).

surrounding. Adiabatic model (18) is extended to three dimensions with heat transfer from lateral surfaces [12–14]:

$$T(x,y,z,t)-T_0=\frac{8q}{c\rho d_L}\sum_{r=1}^{\infty}\sum_{p=1}^{\infty}\sum_{m=1}^{\infty}\frac{(\sin\mu_r)\left(\cos\frac{\mu_r}{d_T}z\right)}{(\mu_r+(\sin\mu_r)(\cos\mu_r))}\frac{(\sin\mu_p)\left(\cos\frac{\mu_p}{d_R}y\right)}{(\mu_p+(\sin\mu_p)(\cos\mu_p))}$$
$$\frac{\mu_m\left(\cos\frac{\mu_m}{d_L}x\right)}{(\mu_m+(\sin\mu_m)(\cos\mu_m))}\left(\frac{1-e^{-\left(\mu_m^2\frac{\alpha_L}{d_L^2}+\mu_p^2\frac{\alpha_R}{d_R^2}+\mu_r^2\frac{\alpha_T}{d_T^2}\right)t}}{\mu_m^2\frac{\alpha_L}{d_L^2}+\mu_p^2\frac{\alpha_R}{d_R^2}+\mu_r^2\frac{\alpha_T}{d_T^2}}\right) \tag{24}$$

where d_L, d_R, d_T are half of dimensions and α_R, α_T, α_L are thermal diffusivities in longitudinal, radial and tangential directions, c is specific heat capacity and ϱ is density at given moisture content. Eq. (24) is the solution of heat conduction Eq. (13), when the block of specimens is in environment of air. The extension for convection at boundaries is accompanied with heat transfer coefficients h and Biot numbers Bi at boundaries. Such extension significantly reduces the number of specimens in quasistationary method to number of 2. Characteristic equations are (for the anatomical direction):

$$\mu_m tg\mu_m=\frac{h_L d_L}{\lambda_L}=Bi_L \tag{25}$$

$$\mu_p tg\mu_p=\frac{h_R d_R}{\lambda_R}=Bi_R \tag{26}$$

$$\mu_r tg\mu_r=\frac{h_T d_T}{\lambda_T}=Bi_T \tag{27}$$

with constant initial temperature through the specimen T_0. The solution (24) fulfils the next boundary conditions:

$$-\lambda_L\frac{\partial T}{\partial x}\bigg|_{x=d_L}=h_L(T|_{x=d_L}-T_0) \tag{28}$$

$$-\lambda_R\frac{\partial T}{\partial x}\bigg|_{y=d_R}=h_R(T|_{y=d_R}-T_0) \tag{29}$$

$$-\lambda_T\frac{\partial T}{\partial x}\bigg|_{z=d_T}=h_T(T|_{z=d_T}-T_0) \tag{30}$$

$$\frac{\partial T}{\partial x}\bigg|_{x=0}=-\frac{\varphi}{\lambda_L} \tag{31}$$

$$\frac{\partial T}{\partial y}\bigg|_{y=0}=0 \tag{32}$$

$$\frac{\partial T}{\partial z}\bigg|_{z=0}=0 \tag{33}$$

where Bi_L, Bi_R, Bi_T are Biot numbers at principal anatomical sections.

The information from only one thermocouple, which is placed in the middle of specimens block, is able to be fully utilized only if material is isotropic in plane of source. Otherwise, more thermocouples must be used to determine all wood thermal properties simultaneously. Another possibility is to rotate the samples considering the position of source and principal anatomical directions. Later all three sets of data will be processed simultaneously. The model (24) is nonlinear. The starting values for least square method are found after utilization of adiabatic models (18) and (23). The solution of heat conduction equation is the base of inverse problem. The solution T_{teor} is compared to measured temperature values at given time T_{exp} in least square criterion Q:

$$Q(c, \alpha_L, \alpha_R, \alpha_T, Bi_L, Bi_R, Bi_T) = \sum_{i=1}^{N} \left(T_{iteor}(c, \alpha_L, \alpha_R, \alpha_T, Bi_L, Bi_R, Bi_T, t_i) - T_{iexp}(t_i) \right)^2 \tag{34}$$

where N is number of measurement.

The results from quasistationary method in three dimensions are summarized in **Tables 1** and **2**.

The beech (*Fagus sylvatica*, L.) and fir (*Abies alba*, Mill.) wood were tested for thermal properties as they are widely used in furniture and construction industry. The 18 beech cubic samples of edge dimension 100 mm in principal anatomical directions were cut from outer part of stem with diameter of 35 cm to be linear orthogonal as much as possible. The specimens' equilibrium moisture content was 12%, and their surfaces were sanded to cube shape of measured dimensions. The 1-cm thick specimen was cut from cubes to place the thermocouple 1 cm far from heating foil. The measurement was performed in climatic chamber with air relative humidity of 65% and temperature of 20°C. The specimens were fixed in the beech rack around the heating foil. The temperature 20°C was initial one and applied heating flux was 145 W m^{-2}. Then temperature was recorded every 5 s. Later on the data in file of recorded times and temperatures were sorted, because long time of results computation according to Eqs. (24) and (34). The experiment was performed three times, each time the heating foil touched the

d_L [m]	0.1082
α_L [m^2 s^{-1}]	2.9×10^{-7}
α_R [m^2 s^{-1}]	1.7×10^{-7}
α_T [m^2 s^{-1}]	1.2×10^{-7}
c [J kg^{-1} K^{-1}]	1900
λ_L [W m^{-1} K^{-1}]	0.38
λ_R [W m^{-1} K^{-1}]	0.23
λ_T [W m^{-1} K^{-1}]	0.16
ρ[kg m^{-3}]	703.9

Table 1. Average thermal properties of beech wood [13].

d_L [m]	0.0542
α_L [m^2 s^{-1}]	5.2×10^{-7}
α_R [m^2 s^{-1}]	2.4×10^{-7}
α_T [m^2 s^{-1}]	1.8×10^{-7}
c [J kg^{-1} K^{-1}]	1700
λ_L [W m^{-1} K^{-1}]	0.36
λ_R [W m^{-1} K^{-1}]	0.17
λ_T [W m^{-1} K^{-1}]	0.13
ρ[kg m^{-3}]	414.5

Table 2. Average thermal properties of fir wood [14].

different section. The results are embedded in **Table 1** and the evaluated temperatures along with computed results are depicted in **Figure 3**.

The same experiment was performed with 10 fir samples. The additional differences were in dimensions (the fir cube edge was 50 mm), position of thermocouple was 3.4 mm next to heating foil and heating flux was 100 W m^{-2}. The results for fir wood samples are in **Table 2** and depicted in **Figure 4**. The convection at boundary was free during both experiments (heat transfer coefficient h_L = 8.9 W m^{-2} K^{-1} at beech cross section and heat transfer coefficient 1.0×10^1 W m^{-2} K^{-1} at fir anatomical sections).

The thermal properties have advantage in common definitions. The solutions of heat conduction equation are expressed in dimensional criterion or numbers. They describe the conduction

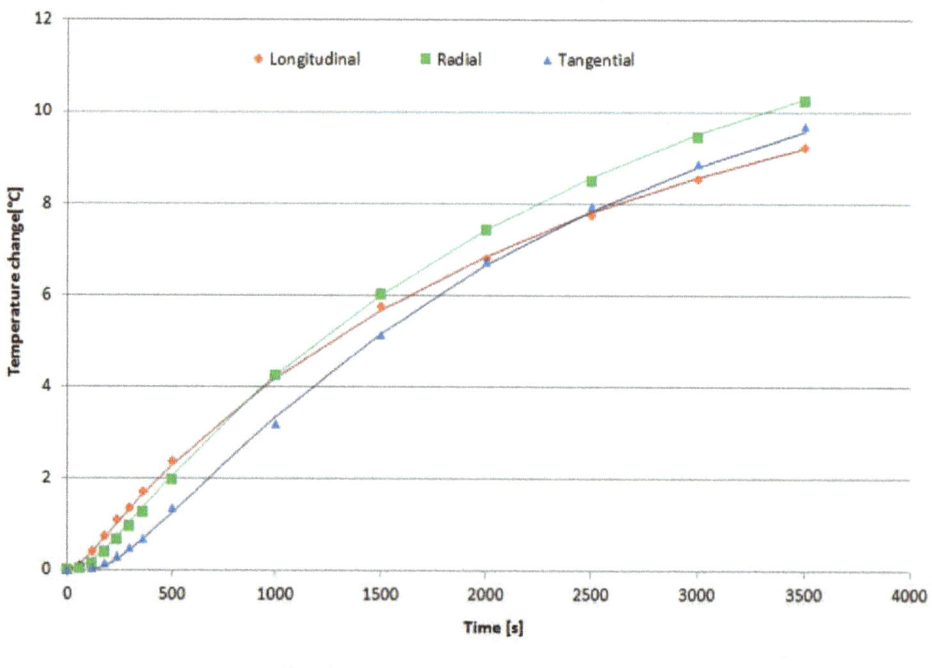

Figure 3. Temperature increase in point [1; 0; 0]cm measured from heating foil placed in different anatomical sections of beech wood. Sum of squares for 42 temperature measurements was 0.21 K^2 [13].

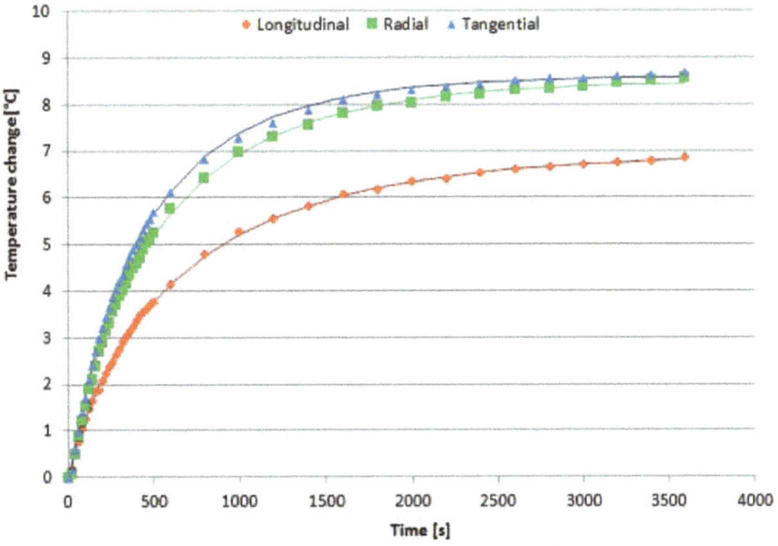

Figure 4. Temperature increase in point [0.34; 0; 0]cm measured from heating foil placed in different anatomical sections of beech wood. Sum of squares for 126 temperature measurements was 0.50K^2 [14].

in similar objects and, moreover, the results can be extended to non-homogeneous objects. Typical example is summing the thermal conductivities in building physics. The result depends on arranging the layers and heat flux direction, **Figure 5**.

Both formulas in **Figure 5** are valid in composite wall at steady state. The following formulas enable the infinitesimal extension to non-homogeneous continuum [7]:

$$\lambda_{\perp} = \frac{d}{\int_0^d \frac{dx}{\lambda(x)}} \tag{35}$$

$$\lambda_{II} = \frac{1}{d}\int_0^d \lambda(x)dx \tag{36}$$

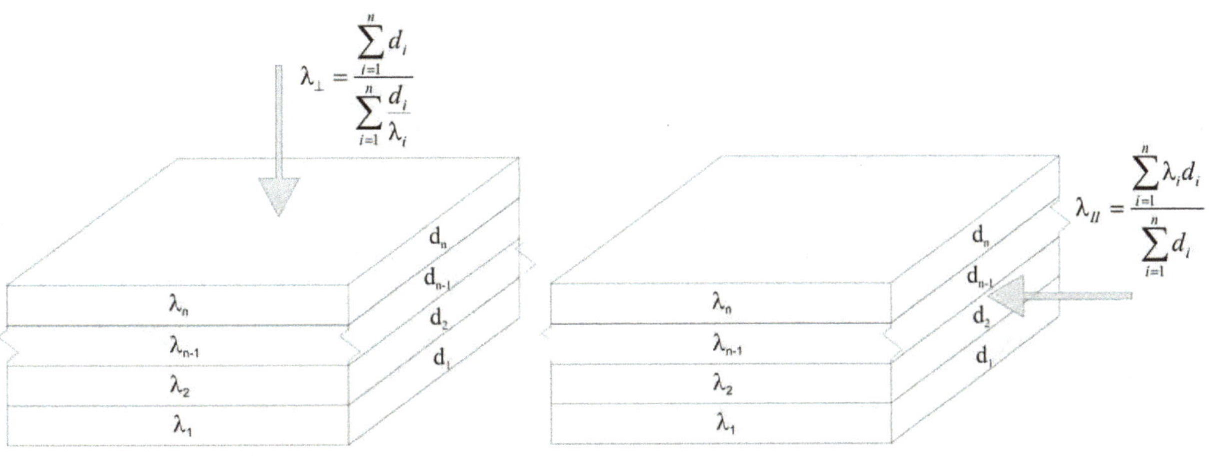

Figure 5. The basic pattern of layers and heat flux (arrow) in composite wall during conduction.

The steady state solution (37) in log with infinite length (for example as element of log cabin houses as is described in Ref. [15], **Figure 6**) is utilized for computing heat flux **q** = q**i** (38) passing through it:

$$T\text{-}T_0 = \frac{1}{2}(T_2\text{-}T_0)\left(1\text{-}\left(\frac{r}{R}\right)^{\sqrt{\frac{\alpha_T}{\alpha_R}}}\cos(\varphi)\right) \tag{37}$$

$$q = \sqrt{\lambda_R \lambda_T}\frac{T_2\text{-}T_0}{2R} \tag{38}$$

The "enlarge cracks" are not necessary present in log after conditioning, but regularly one occur after kerfing of logs from sap to pith because shrinkage differences in various anatomical directions.

One of the haptic phenomena—tactile warm—is related to thermal properties [16]. Touching wood at cold winter or hot summer is more pleasant than touching many other materials. One of the possible reason and next explanation of tactile warms as physiological event inheres in value of formed temperature at the surface between wood and human (living) body [17]. The formed temperature is closer to temperature of human body before touching wood either in cold winter or hot summer. The quantity responsible to this event is called thermal effusivity e:

$$\mathrm{e} = \sqrt{\lambda c \rho} \tag{39}$$

as a square root of product of thermal conductivity, specific heat and density. Let two semi-infinite solids have different initial temperatures, T_{01}, T_{02} and no additional sources or sinks act at the surface or in their volumes. Steady temperature T_S at the surface between these two semi-infinite solids in contact is influenced by their effusivities e_1, e_2:

$$\mathrm{T_s = T_{02} + (T_{01}\text{-}T_{02})\frac{e_1}{e_1+e_2} = T_{01} + (T_{02}\text{-}T_{01})\frac{e_2}{e_1+e_2}}. \tag{40}$$

The effusivity values for previously mentioned beech and fir wood are in **Table 3**.

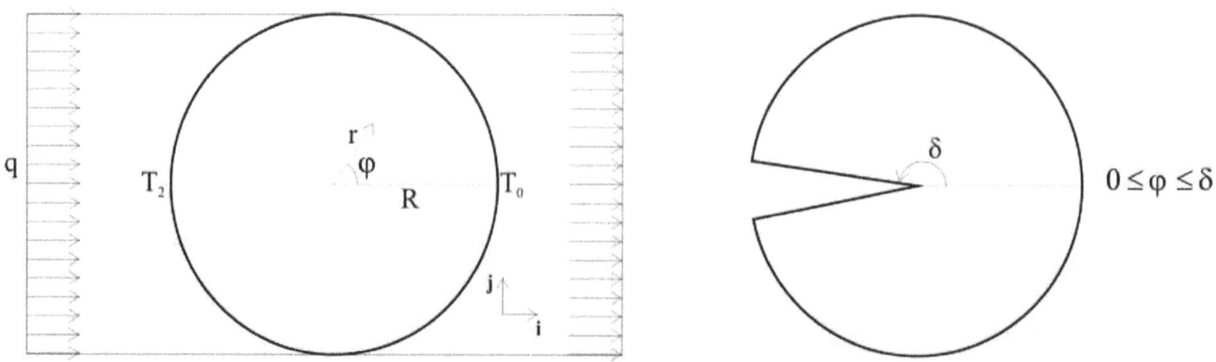

Figure 6. Orientation of heat flux through log (cross section at left side); orientation of heat flux and position of crack due to anisotropy of shrinkage (cross section at right side).

Material	Direction		
	e_L [W m^{-2} K^{-1} s$^{0.5}$]	e_R [W m^{-2} K^{-1} s$^{0.5}$]	e_T [W m^{-2} K^{-1} s$^{0.5}$]
Beech	7.1×10^2	5.5×10^2	4.6×10^2
Fir	5.0×10^2	3.5×10^2	3.0×10^2

Table 3. Thermal effusivity of beech and fir wood in principal anatomical directions.

The largest thermal effusivity eigenvalue is in wood longitudinal direction. It results to the larger drop value of human skin temperature or smaller drop value of wood surface temperature. The opposite is valid for wood tangential direction.

3. Factors related to thermal properties

The most of the factors are quantities: density, moisture content and temperature which are scalars. The dependency on anatomical direction—the direction of measurement, is crucial for distinguishing the thermal property as scalar or tensor.

3.1. Anatomical direction

Specific heat is a scalar quantity neither it is mass nor volume specific heat capacity, they do not depend on anatomical direction. Thermal conductivity and thermal diffusivity are second order tensors. Their eigendirections coincide with principal anatomical directions as a result of measurement (also proved by Sonderegger et al. [18], Vay et al. [19]). Also their eigenvalues determine the dimensions [$2.d_{Rmin}$, $2.d_{Tmin}$, $2.d_{Lmin}$] of the parallelepiped of the slowest average temperature change in its volume V:

$$d^3_{Rmin} = \frac{V}{8}\frac{\alpha_R}{\sqrt{\alpha_T\alpha_L}}, \quad d^3_{Tmin} = \frac{V}{8}\frac{\alpha_T}{\sqrt{\alpha_L\alpha_R}}, \quad d^3_{Lmin} = \frac{V}{8}\frac{\alpha_L}{\sqrt{\alpha_R\alpha_T}} \tag{41}$$

Then, the ratio of optimal dimensions for beech in longitudinal, radial and tangential directions is:

1.6:1.2:1

and for fir:

1.7:1.2:1.

The characteristic feature of the optimal parallelepiped is the smallest transferred heat through its surfaces, for example, from its volume.

3.2. Density

Wood is regarded as non-homogeneous material. This conclusion is strongly supported by microscopy [20]. Therefore, for every homogeneous part of wood volume, we should prescribe the equations with appropriate boundary conditions. The phenomenon of continuum for whole wood volume has been more efficient yet. We often neglect mass of air inside wood as it is measured in air environment. The air volume in wood is not neglected. Therefore, mass specific heat capacity c_0 does not depend on oven dry density or anatomical species. Many experiments proved the significant thermal conductivity dependency on oven dry density. Wood contains air as a good thermal insulator, therefore, a smaller value of oven dry density results in a smaller value of wood thermal conductivity. Perhaps one of the first relationships was set in theory of thermal bridges by Kollmann and Malmquist [21]. The theory utilizes the equations in **Figure 5**, and oven dry wood is treated as composites of substance and air. Also, the theory stated the dependency of thermal conductivity on anatomical directions according to value of bridge factor. In spite of thermal bridge theory indisputability, it seems to be more efficient the relationship between thermal conductivity and density at given moisture content. If density is zero then no matter exists for conduction and the simplest non-homogeneous model (layer of wood substance next to layer of air) predicts linear relationship between thermal conductivity and density at given moisture content and later no influence of density at given moisture content on thermal diffusivity. These results are in contrary to results in **Tables 1** and **2**, so the simplest non-homogeneous model does not hold for wood well. Week relationship between thermal diffusivity and density at given moisture content was published by Harada et al. [22].

3.3. Moisture content

The ratio of water mass specific heat to air mass specific heat is more than one. It follows from mixing rule [23]:

$$c = \frac{c_0 + wc_{H_2O}}{1+w} - \frac{1}{1+w}\frac{d(Q_{z0} - Q_z)}{M_{H_2O}dT} \tag{42}$$

where Q_z, resp. Q_{z0} is wetting heat at given moisture content, resp. in oven dry state and is represented in J mol^{-1}. The last expression of Eq. (42) is zero for free water. The water molecules in wood behave as sinks during evaporation. It should be noted, that diffusion of water in wood is approximately 100 times slower than conduction of heat through wood at room temperatures and these two processes can be studied separately [7]. The relationship between equilibrium moisture content and species is weak. Also, the influences of moisture content and oven dry density on thermal conductivity and thermal diffusivity are studied separately. The density at given moisture content is directly connected with thermal diffusivity computing formula. Then, the influence of moisture content on thermal diffusivity is set into density at given moisture content.

3.4. Temperature

The linear function of mass specific heat capacity at zero moisture content on temperature was provided by Perelygin [24] (according to Požgaj et al. [7]):

$$c_0 = 1.571 + 0.00277\vartheta \ [\mathrm{kJkg^{-1}K^{-1}}] \tag{43}$$

and [25] (according to Radmanović et al. [26]):

$$c_0 = 1.1136 + 0.004856\vartheta \ [\mathrm{kJkg^{-1}K^{-1}}] \tag{44}$$

and others in the range ϑ = 0–100°C, even [27] published theoretical function determined from Debye's theory for solid substances at constant volume. The advantage of linear relationship is in easy computation of average specific heat value. Also, determination of adiabatic temperature leads to quadratic equation. The influence of temperature on thermal conductivity is studied by setting initial temperatures to different levels and only small temperature drop between surroundings and tested specimen causes heat transfer. Then results are tested by regression even the results were obtained with method using the heat conduction equation solution with constant coefficients. Later, on such results are used in numerical inverse or direct problem solutions as starting values. The solving of nonlinear equation is better solution to incorporate the dependence of thermal properties on temperature. One of the possibilities to overcome this problem is designing the dependency of thermal conductivity on temperature, for example as polynomial:

$$\lambda = k(T - T_\infty)^n \tag{45}$$

where T_∞, k, n are constants and heat conduction equation is rearranged to the parabolic equation again:

$$\frac{\partial F}{\partial t} = \alpha \frac{\partial^2 F}{\partial x^2} \tag{46}$$

where

$$F = \frac{k(T - T_\infty)^{n+1}}{n+1} \tag{47}$$

Also, the Arhenius dependence of thermal conductivity on temperature leads to Eq. (46), but transformation from F to T is nonlinear and must be find numerically. The numerical solutions of nonlinear heat conduction equation were showed by Zhao et al. [28]. Their procedure was applied on experimental data of varying temperature and moisture content according to industrial practise.

3.5. Thermal expansion

The change of heat capacity at constant pressure C_p and at constant volume C_V is distinct for gases. This phenomenon is often neglected for wood even the coefficient of volume thermal expansion α_e is determined:

$$V - V_0 = \gamma_e V_0 \vartheta \tag{48}$$

where V is volume at temperature ϑ in °C, V_0 is the volume at 0°C. Then, difference in capacities is:

$$C_p - C_V = TV\frac{\gamma^2}{K_T} \tag{49}$$

where T is temperature in K and K_T is isothermal compressibility. The coefficient of linear thermal expansion α_e is defined:

Material	Property			
	$10^6 \times \alpha_{eT}$ [°C^{-1}]	$10^6 \times \alpha_{eR}$ [°C^{-1}]	$10^6 \times \alpha_{eL}$ [°C^{-1}]	$10^6 \times \gamma_e$ [°C^{-1}]
Fir[a]	31.6	21.7	3.90	57.2
Beech[b]	40.3	31.8	3.1–4.5	75.2–76.6

Table 4. The coefficient of linear thermal expansion of fir and beech wood at zero moisture content ([29][a], [30][b]) and coefficient of volume thermal expansion.

$$d{-}d_0 = \alpha_e d_0 \vartheta \tag{50}$$

where d is dimension at temperature ϑ in °C, d_0 is the dimension at 0°C. The relationship between coefficients of linear and volume thermal expansion is:

$$\gamma = \alpha_{eL} + \alpha_{eR} + \alpha_{eT} + \alpha_{eL}\alpha_{eR} + \alpha_{eR}\alpha_{eT} + \alpha_{eL}\alpha_{eR} + \alpha_{eL}\alpha_{eR}\alpha_{eT} \approx \alpha_{eL} + \alpha_{eR} + \alpha_{eT} \tag{51}$$

where L, R, and T denotes principal anatomical directions (**Table 4**).

The dimension of wood equilibrates as fast as temperature of wood equilibrates. And thermal diffusivity is also property characterizing how fast the dimensions of wood equilibrate during equilibrating its two different temperatures. Let it say, for temperature T in point x_0 and time t is valid:

$$T(x_0, t) = T_\infty + (T_0 - T_\infty)e^{-\frac{h}{d_0}t} \tag{52}$$

with no-zero flux at the surface $x_0 = d_0$, where T_∞ is equilibrium temperature and T_0 is initial temperature. If conduction occurs during constant pressure, the new position of point x_0 will be coordinate x:

$$x = x_0\left(1 + \alpha_e(T_0 - T_\infty)e^{-\frac{h}{d_0}t}\right) \tag{53}$$

Two dimensional problems are more complicated, but bonded oven dry specimens with thin thicknesses oriented in different anatomical directions and different initial temperatures show distinct deflections due to anisotropy of coefficient of linear thermal expansion in equilibrium.

4. Conclusion

The answers to the three questions mentioned in Section 1 are closely related to thermal properties. Wood as natural material is characterized by its properties, which definitions are precisely stated. The definitions are often expressed as equations which solutions are used in measuring methods. Because of many measuring methods of wood thermal properties and large variability of their results, methods must be clearly explained and their technical representation must be closed as much as possible to assumptions used in solutions. The large variability of the results is connected not even with inherent categorization of wood species, but for example also position in steam causes variability of the results. There are several factors which affect the wood thermal properties in wood and a lot of others waited for discovering [31]. The management of the industrial processes in real time according to properties are really difficult. The designs of the heat transfer processes are connected to wood thermal properties. After measuring of the results, they are used in solutions of different similar direct problems, for example in furniture design or building physics and others.

Author details

Richard Hrčka* and Marián Babiak

*Address all correspondence to: hrcka@tuzvo.sk

Department of Wood Science at Technical University in Zvolen, Zvolen, Slovak Republic

References

[1] Goodstain D. Thermal physics. 1st ed. United Kingdom: Cambridge University Press; 2015. 165 p.

[2] Babiak M. Niektoré tepelné vlastnosti smrekového dreva a z neho vyrobených materiálov [dissertation]. Some thermal properties of spruce wood and materials made of it. Zvolen: Vysoká škola lesnícka a drevárska, Drevárska fakulta; 1976.

[3] Khattabi A, Steinhagen HP. Analysis of transient nonlinear heat conduction in wood using finite-difference solutions. Holz als Roh- und Werkstoff. 1993;**51**:272–278.

[4] Deliiski N, Dzurenda L, Miltchev R. Computation and 3D visualization of the transient temperature distribution in logs during steaming. Acta Facult. Xylol. Zvolen. 2010;**52** (2):23–31.

[5] Lykov AV. Analytical heat diffusion theory. 2nd ed. New York: Academic Press, Inc.; 1968. 685 p.

[6] Logan DJ. Applied mathematics. 4th ed. Hoboken, New Jersey: John Wiley & Sons, Inc.; 2013. 658 p.

[7] Požgaj A, Chovanec D, Kurjatko S, Babiak M. Štruktúra a vlastnosti dreva. Wood structure and properties. 1st ed. Bratislava: Príroda; 1993. 486 p.

[8] Babiak M, Krakovský A. The behavior of heated wood loaded in bending. In: Wood and Fire Safety; Technical University of Zvolen. Technical University of Zvolen; 2000. 9–16 p.

[9] Clarke LN, Kingston RST. Equipment for the simultaneous determination of thermal conductivity and diffusivity of insulating materials using a variable state method. Aust. J. Appl. Sci. 1950;**1**:172–187.

[10] Babiak M, Hrčka R. Measurement of wood thermal properties. In: Morén T, Antti L, Sehlstedt-Persson M, editors. Recent advances in the field of wood drying; January 18–22, 2010; Skelleftea. Sweden: LTU Skelleftea; 2010. 415 p.

[11] Krischer O, Esdorn H. Einfaches Kurzzeitverfahren zur gleichzeitigen Bestimmung der Wärmeleitzahl, der Wärmekapazität und der Wärmeeindringzahl fester Stoffe. Simple short-term method for the simultaneous determination of heat conductivity, heat capacity and heat diffusivity of solids. VDI Forschungsheft. 1954;**450**:28–39.

[12] Hrčka R.. Variation of thermal properties of beech wood in the radial direction with moisture content and density. In: Kúdela J, Lagaňa R, editors. Wood Structure and Properties '10; September 6–9, 2010; Podbanské, High Tatras. Zvolen, Slovakia: Arbora Publishers; 2010. 111–115 p.

[13] Hrčka R, Halachan P. Some measurement of wood thermal properties. In: Researchers & producers V-4: development of cooperation in the field of passive and zero buildings. In: Pasztory Z, Ihnát V, editors. Developement of cooperation in the field of passsive and zero buildings; May 16–17, 2013; Sopron, Hungary. Bratislava: Pulp and Paper Research Institute; 2013. 23–27 p.

[14] Hrčka R, Babiak M. Meranie tepelných vlastností dreva. Measurement of wood thermal properties. In: Dřevostavby, dřevo, surovina moderního člověka; April 1–2, 2016; VOŠ Volyně. Czech Republic : VOŠ Volyně; 2015. 24–28 p.

[15] Jochim S. Determining the heat transfer coefficient of log-cabin walls based on one dimensional thermal transmittance. Acta Facult. Xylol. Zvolen. 2016;**58**(1):75–82. doi:10.17423/afx.2016.58.1.0

[16] Haviarová E, Babiak M, Nemec Ľ, Joščák P. Teplotný vnem človeka a tepelné vlastnosti vybraných materiálov a drevín používaných k tvorbe komplexného interiéru. Temperature sense of a person and thermal properties of selected materials and wood species used at the complex interior formation. Acta Facult. Xylol. Zvolen. 1996;**38**(2):7–14.

[17] Novotný R, Kúdela J, Hrčka R, Lagaňa R, Moravec I. Multimedial presentation as a new tool for interpretation and communication in area of wood science. Acta Facult. Xylol. Zvolen. 2010;**52**(2):5–15.

[18] Sonderegger W, Hering S, Niemz P. Thermal behaviour of Norway spruce and European beech in and between the principal anatomical directions. Holzforschung. 2011;**65**:369–375. doi:10.1515/HF.2011.036

[19] Vay O, De Borst K, Hansmann Ch, Teischinger A, Muller U. Thermal conductivity of wood at angles to the principal anatomical directions. Wood Sci. Technol. 2015;**49**:577–589. doi:10.1007/s00226-015-0716-x

[20] Hunt JF, Gu H, Lebow P. Theoretical thermal conductivity equation for uniform density wood cells. Wood Fiber Sci. 2008;**40**(2):167–180.

[21] Kollmann F, Malmquist L. Über die Wärmeleitzahl von Holz und Holzwerkstoffen. Holz Roh. Werkst. On the thermal conductivity of wood and wood – based materials. 1956;**14**:201–204.

[22] Harada T, Hata T, Ishihara S. Thermal constants of wood during the heating process measured with the laser flash method. J Wood Sci. 1998;**44**:425–431.

[23] Hearmon RFS, Burcham JN. The specific heat and heat of wetting of wood. Chem. Ind. 1956;807–808.

[24] Perelygin LM. Náuka o dreve. Wood science. 2nd ed. Bratislava: SNTL; 1965. 382 p.

[25] Dunlop F. The specific heat of wood. US Dept. Agric. For. Serv. Bull. 110; 1912. 28 p.

[26] Radmanović K, Đukić I, Pervan S. Specific heat capacity of wood. Drvna Ind. 2014;**65**(2):151–157. doi:10.5552/drind.2014.1333

[27] Ward RJ, Skaar Ch. Specific heat and conductivity of particleboard as functions of temperature. For. Prod. J. 1963;**1**:31–38.

[28] Zhao J, Fu Z, Jia X, Cai Y. Inverse determination of thermal conductivity in lumber based on genetic algorithms. Holzforschung. 2016;**70**(3):235–241. doi:10.1515/hf-2015-0019

[29] Kollmann F, Côte WA. Principles of wood science and technology. 1st ed. Berlin-Heidelberg-New York: Springer-Verlag; 1968. 592 p.

[30] Glass SV, Zelinka SL. Moisture relations and physical properties of wood. In: Wood handbook, Wood as an engeneering material. Madison, Wisconsin: USDA Forest Service; 2010. 1–19 p.

[31] Hrčka R, Babiak M. Some non-traditional factors influencing thermal properties of wood. Wood Res. 2012;**57**(3):367–373.

6

Traditional Wooden Buildings in China

Ze-li Que, Zhe-rui Li, Xiao-lan Zhang, Zi-ye Yuan and Biao Pan

Abstract

Chinese ancient architecture, with its long history, unique systematic features and wide-spread employment as well as its abundant heritages, is a valuable legacy of the whole world. Due to the particularity of the material and structure of Chinese ancient architecture, relatively research results are mostly published in Chinese, which limits international communication. On account of the studies carried out in Nanjing Forestry University and many other universities and teams, this chapter emphatically introduces the development, structural evolution and preservation of traditional Chinese wooden structure; research status focuses on material properties, decay pattern, anti-seismic performance and corresponding conservation and reinforcement technologies of the main load-bearing members in traditional Chinese wooden structure.

Keywords: traditional Chinese wooden structure, materials and properties, anti-seismic performance, reinforcement techniques

1. Introduction

Being one of the world's three major architecture systems, Chinese ancient architecture plays an important role in the global history of architecture. With its long history, unique systematic features and wide-spread employment as well as its abundant heritages, Chinese ancient architecture keeps growing and developing. Emerging from a system using earth and wood to one using bricks and wood, it held on its tradition of taking wooden structure as the main structure and carpentry as the main technology. After over 2000 years of progression and evolution, it has formed a complete system of structure and construction, which includes regulations and standards inherited both from Song Dynasty (1103 AD) and from Qing Dynasty (1734 AD). Compared to Western ancient buildings constructed with stones, bricks and natural concrete, Chinese traditional wooden buildings lack durability and need frequent maintenance and renovation, and properties of the wood in use have a fairly big influence on the joints and the performance of the whole structure. However, under the influence of traditional Chinese philosophy, buildings have been more of an exhibition of social status and the materials and structures involved have not been taken seriously as a technology for a long time.

The study with significance to the modern world in the field of Chinese traditional wooden buildings started in the 1920s and 1930s. Historic and artistic fields of the architecture attracted most attention and were often selected as main research directions over a long period of time. Up to now, limited number of fundamental studies on the structural behaviour of Chinese traditional timber structure and its typical joint connections can be found; hence there is an urgency to study and evaluate the seismic performance and structural behaviour of the existing historical timber buildings so as to prevent as much earthquake-inflicted damages as possible from occurring in the near future.

Taking Dou-gong brackets and mortise and tenon joints of Chinese traditional timber structure as objects, the ongoing research project of our team includes structure performance and anti-seismic mechanism of different joint connections between columns and beams, reinforcement technology of the weak parts, along with the utilization and analysis of modern engineering wood products as alternative materials in the repair and new construction of Chinese traditional timber constructions.

Material performance and structure behaviour researches of Chinese traditional wooden buildings are often based on specific emergency repairment and strengthening projects of historical buildings, which somewhat limits the systematicness and universality of the researches. On the other hand, taking convenience of cultural awareness and characteristic of oriental structural system into consideration, the results of relevant studies tend to be published domestically, which also increases the difficulty of international academic exchange and interaction. In consequence, the intention of this chapter is to collect and introduce relevant research status as well as phased achievements of my team systemically. And the publication of this book will be certain to generate a trend to study traditional wooden structure and encourage worldwide academic exchange and cooperation.

2. The structure and preservation of traditional Chinese wooden architecture

Represented by traditional Chinese wooden architecture, oriental wooden structure stands out in the architecture world, and after a long course of development and accretion, it has reached a high level of standard theoretically and practically. Take the example of the Yingxian Wooden Pagoda, the highest wooden tower existing worldwide. Besides the fact of being 67.1-m high, it has also survived several major earthquakes and therefore embodied the perfect combination of techniques and aesthetics of wooden structures as well as the intelligence of ancient Chinese people. Consulting two significant building standards from Song Dynasty and Qing Dynasty, this chapter introduces the development and structural evolution of traditional Chinese wooden structure, focusing on three classic structures and via the examples of well-known wooden structures such as the Yingxian Wooden Pagoda, and presents the condition of study and preservation of historic buildings in modern China.

2.1. A brief guide to the evolution of traditional oriental wooden structures

Due to different cultural backgrounds, ancient architecture used to have seven independent systems, of which some are extinct or never widely spread and thus had limited achievements and influences. That left Chinese architecture, European architecture and Islamic architecture to be considered the world's three main architectural systems. And among them, Chinese architecture and European architecture are the most long-lasting, widely spread and successful ones. Ancient Chinese architecture had undergone primitive society, slave society and feudal society, among which the last one was the time when Chinese classic architecture developed the most.

1. **Primitive society (7000 years ago to twenty-first century BC)**. The building types vary due to different climates, geographical features and materials. Among them, there are two typical types: wooden frame and mud wall buildings that emerged from cave houses in the Yellow River basin and the Ganlan-style buildings (wooden buildings that built on stilts) from nest houses in the Yangtze River basin. In the late stage of the primitive society, building sites already had trace of privatization and the walls and roofs of buildings were mostly interwoven branches or twigs with mud coating (see Ref. [1]).

2. **Slave society (2070–476 BC)**. In the twenty-first century BC, the wooden frame and rammed earth construction and regular enclosed courtyard building groups came along, which showed great improvement in timber frame technology. The sixteenth century BC was the prime time for the development of the Chinese slave society and a time when documentary trace began. Based on the size of the rammed earth foundation of the palaces and temples, buildings at this point of the history had larger scale and stricter hierarchy and scale of cities, height of city walls, width of streets and other buildings of significance were required to be built according to their rank. In the Spring and Autumn periods (770–476 BC), the popularization of tiles and appearance of high-platform buildings for imperial and ducal palaces were the most important improvements. High-platform building means building a platform of tamped earth underneath the palace. As the leuds sought more magnificent palaces, the decoration and painting of ancient architecture were taken a step further (see Ref. [1]).

3. **Feudal society (475 BC to 1911 AD)**. With the collapse of slavery, agriculture and handicraft rapidly grew and the utilization of ironware accelerated the improvement of structure technology and wooden structure's construction quality. Fireplaces, heated brick beds and cellars can be seen at this period of time. The Han Dynasty was a thriving time for classic Chinese architecture when the nowadays commonly seen beam-lifted frame and through-type frame wooden structures were formed. And at the same time, the traditional roof of Chinese buildings also flourished. Since then, the introduction of Buddhism greatly boosted the development of Buddhist architecture, one of the most

important types of classic Chinese architecture. The Tang Dynasty was a time when the techniques and artistic qualities of classic architect were developing the fastest. Tang-style architecture demonstrates the extremity of size and regulations, extremity in architectural complex layout and features of large expansion and large volume. And the construction form and material requirements of wooden structures especially Dou-gong brackets were standardized. Tang-style architecture also produced a far-reaching influence on countries such as Japan. Later in the Song Dynasty, modular system was adopted and the book building standards were officially published which set standard rule for buildings' measurements and basic moduli so that the size of wooden components could be properly defined.

In the late stage of feudal society, building forms were becoming more and more simplified and the entirety of the beam-column frame was enhanced. The buildings presented a serious and rigorous image with more ingrained decoration and painting. In Qing Dynasty, the ethnic diversity contributed to the blossoming of various residential building types. And the monomer building form of official architecture was set and therefore improved the standard of architectural complex design. The promulgated book construction practices enumerated 27 practices of monomer building and formulated new construction moduli, which contributed much to accelerate the design and construction process and controlling material consumption (see Ref. [2]).

2.2. The structural system and characteristics of traditional Chinese wooden structure

Based on different construction frames and geographic features, the traditional Chinese wooden structure frame system can be divided into three types: through-type frame, beam-lifted frame and log-cabin-type frame (see Refs. [3, 4]), as seen in **Figure 1**.

1. **Through-type frame**. The through-type frame is constructed of vertical connection with separated frame and mostly used in rural housing. There is no reference to this type in the official building standards. The common practice of this type is to connect the columns with square crossbeams along the length of the house, forming a truss and then use square crossbeams to connect every two trusses, forming the frame of the house. The characteristics of this type include using materials with small cross section that are easy to obtain, using multiple square crossbeams along the length of the house that can be assembled beforehand, enhancing the entirety and stability of the structure and ren-

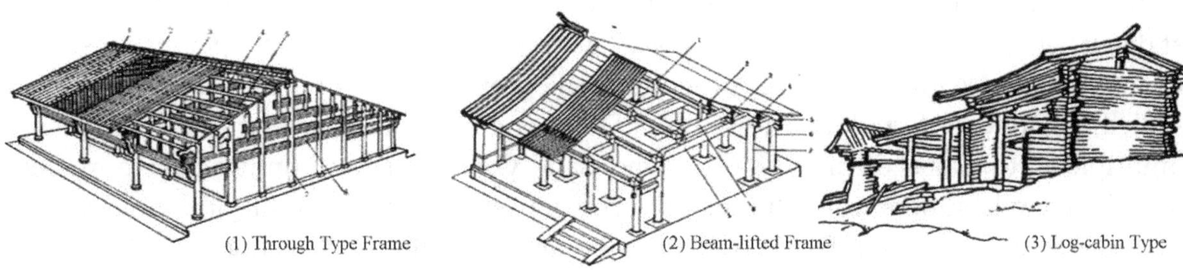

Figure 1. Three types of traditional Chinese wooden structure.

dering the installation of walls convenient and saving manpower and materials with its simple practice, direct force transmission and ever-evolving and adaptable nature.

2. **Beam-lifted frame**. This frame type formed in the Spring and Autumn period kept evolving and then became a settled practice. This frame type varies in material size and frame combination according to different social ranks, which was strictly set in regulations such as building standards in Song Dynasty and construction practices in Qing Dynasty. Beam-lifted frame is usually composed of the frame layer, the Dou-gong brackets layer and the roof layer. Usually, it is constructed by placing a beam head on top of a column and then on top of that, using a shorter column to hold a shorter beam and another beam head and another column and so forth. And eventually, the short column on the short beam holds the weight of the purlin. This type was widely used in large-scale buildings such as palaces and temples in northern China. The characteristics of beam-lifted frame are long distance between columns along the length of the house, enclosing larger interior space and aesthetically pleasing structural features.

3. **Log-cabin type**. Log-cabin type is an ancient structural type that dates back to the primitive society. In China, it was found to be used in building the outer coffin in Shang Dynasty tombs from over 3000 years ago and in the caved patterns on Han Dynasty relics found in Yunnan province in south-western China. It is referred as 'Mukeden' in north-eastern China, meaning to pile up caved logs (often cut into semi-cylinders) to build houses. This type of structures is often seen in areas such as Inner Mongolia, forests in north-eastern China and mountain areas in Sichuan province and Yunnan province in south-western China. Its characteristics are as follows: it can regulate the room temperature to fit the fickle climate in mountain areas and can withstand earthquakes to some extent; it requires only simple materials and minimum manpower but possesses great diversity and mobility; however, to build this type of houses, a great amount of wood is required and the size and location of doors and windows are greatly limited so it is not as widely spread as the other two types.

2.3. The preservation and research status of two typical remaining historic wooden buildings in China

(1) Yingxian Wooden Pagoda. Yingxian Wooden Pagoda, originally known as the Yingxian Wooden Pagoda of Fogong Temple, was built in 1056 AD, Liao Dynasty, and is the largest and oldest high-rise wooden building in existence in the world (as seen in **Figure 2**). It is a 67.31-m tall pagoda of a multi-storied pavilion type with an octagonal cross section and nine floors that disguised as five. It has a diameter of 30.27 m, weighs 7400 tons and altogether consumed 3700 m^3 of timber. With 54 types of Dou-gong brackets of different functions, shapes and sizes installed, it is often referred as a museum of Dou-gong bracket. However, as a consequence of multiple earthquakes during the recent thousand years, wars and unfit repair in modern times, the pagoda suffers from all kinds of problems such as a severe tilt of the main body and the twist of the column frame of the second and third floors. Based on the observation data of 2010, the overall slope was 1.25% and counting, especially of the second floor which accounted for 60–70% of the slope.

Figure 2. Yingxian wooden pagoda.

Since 1933, Liang et al. began to conduct detailed researches and measurement on Yingxian Wooden Pagoda. In 1966, the book Yingxian Wooden Pagoda was published, and in 1973 (see Ref. [5]), architectural experts such as Yang Tingbao began their 10 years of restoration of this architectural treasure after discussing the issue of its partial tilt and setting basic rules and solutions regarding the repair and reinforcement of the pagoda. The Committee of Yingxian Wooden Pagoda Restoration and Preservation Construction Management was found in the 1990s, and after the early-stage study, it started monitoring the structural soundness of the pagoda in 2008 and continued till this day. Since the 1990s, many scholars and their teams have studied the structural state (e.g. Ref. [6]), damage dispersion, seismic reaction analysis (e.g. Ref. [7]) and material deformation (e.g. Ref. [8]) features under external forces. Refined finite element (FE) models were established, respectively, based on the Dou-gong bracket joints and the whole structural system and load-bearing quality analyses were conducted under lateral load (e.g. Ref. [9]). And the ideal restoration model of the pagoda was established through computer-aided design (CAD) drawings and three-dimensional (3D) models (see Ref. [10]). Yet, there are still issues to address in terms of repair and preservation. In recent years, scholars came up with plans such as major repair of the framework, total support of the pagoda and raised support of the upper section. But because of the significance, structural complexity and uniqueness of the pagoda, the present plan is to reinforce and repair the tilted parts and damaged components on the second and third floors (see Ref. [11]).

(2) East palace of Foguang Temple. The palace is located in Wutai county, Shanxi province, in northern China and originally built in Northern Wei Dynasty (386–534 AD, one of the

Northern Dynasties), as seen in **Figure 3**. With the remaining main hall rebuilt in 857 AD, Tang Dynasty, the palace is one of the remaining oldest Tong Dynasty wooden structures and acclaimed as 'the primary national treasure of China'. With a building width of seven rooms and length of four, the roof, column frame and Dou-gong brackets all belong to the top rank and exhibit classic structural features of Tong Dynasty. The Dou-gong components have cross-sectional size of 210 × 300 cm, 10 times the size of the same type of components in Qing Dynasty. The eaves are 3.69 m long, and the triangle Y-shape support system in the beam frame is the first of its kind in China. In the palace, there are 61 m^2 of Tang Dynasty wall paintings preserved and other treasures such as inscriptions from Tong Dynasty and painted sculptures.

As a classic example of the structural frame and construction technology of Tang Dynasty architecture, multi-dimensional studies were carried out on the East Palace of Foguang Temple regarding spatial form, structural bearing capacity, anti-seismic reinforcement and artistic characteristics (e.g. [12, 13]). As to the protection of the palace, surrounding residents were moved out in 1954 and repair and reinforcement began. In 1985, local government built dams around it to protect it from mountain torrents and stone walls, flashing and gutters to reduce humidity.

Figure 3. East palace of Foguang temple.

3. Properties study on traditional Chinese wooden structure

3.1. Study of physical properties on wood materials from historic buildings

In order to protect all the wooden buildings with hundreds of years of history across China, a research team was formed to carry out field study on the structure materials of historic buildings in 11 provinces, municipalities and autonomous regions. Experiments were carried out on the worn components after replacement and thus conclusions were drawn concerning the species and physical properties. Furthermore, a national mandatory standard *Technical code for maintenance and strengthening of ancient timber buildings*, see Ref. [14], was established.

3.1.1. Researches of the species of wooden materials used in historic buildings

According to the field researches and some microstructure identifications, the main components in historic buildings such as beam, fang (a square pillar), column, purlin and rafter are mainly made of nanmu (*Phoebe zhennan S. Lee*), cypress (*Cupressus funebris Endl.*), China fir (*Cunninghamia lanceolata (Lamb.) Hook.*) and pinus (*Pinus massoniana Lamb.*) in southern China. In northern China, Chinese pine (*P. tabulaeformis Carr.*), larch (*Larix gmelinii (Rupr.) Kuzen.*) and pinus armandi (*P. armandii Franch.*) are widely used. And in common housing, poplar (*PopulusL.*) and elm (*Ulmus pumila L.*) are used while in large-scale historic buildings of great importance, there are wood materials such as nanmu from the south involved, which indicates that these buildings are mostly constructed with local materials except for important ones with higher standards. Buildings in Sichuan province in southwest China and Hubei province in the middle south are constructed of nanmu and cypress for there was a large reserve at the time. However, the value of this wood goes up as the reserve goes down nowadays.

3.1.2. Studies on the physical properties of the components in historic buildings

The first concern of all the people working in the field of ancient architecture is the change pattern of the properties of load-bearing components. But studies concerning this topic or relevant ones have been hard to find. The leading difficulty lies in that the environmental conditions that the components are in play an important role in how their properties change and different species react largely differently to the environmental conditions. In addition, it is also difficult to manufacture viable specimens using modern materials as a control group due to the massive variation of wood materials.

In 1977, Chen G.Y carried out physical properties experiments on a worn component from the Yingxian Wooden Pagoda, see Ref. [15]. It was a column on the horizontal slot of the two-raftered roof beam on the second floor that was 900 years old according to C14 dating. The column was 2.7-m high, 33 × 23 cm in section size and made of north China larch (*L. principis-rupprechtii Mayr*). Being hidden inside the pagoda, the column was spared from erosion by wind and rain and thus showed no obvious erosional furrows and darkening but demonstrated some split (being hit by artillery shells). The experiment results are seen in **Table 1**. In 1982, Chen experimented on the middle column from the Jing Qing Gate in Jinci Temple, see Ref. [15]. The column was about 600 years old, 6-m high, 40 × 40 cm in section size, and made of poplar (*Populus L.*). The column was not eroded by rain and demonstrated darkening and different levels of split. It showed trace of weathering and was eroded into powder at approximately 1 m above the root, rendering the root conical and leaving the upper half relatively intact. The results are seen in **Table 2**.

Both experiments proved that after 600–900 years of load bearing, the tensile strength and compressive strength perpendicular to the grain of the material were weakened the most:

the former by 50% and the latter by 80% in pinus and poplar, respectively. At the same time, the stiffness and shear strength were enhanced: the former by 11–16% and the latter by 15%. This indicates that old wood material has denser cell structure and therefore higher level of stiffness than new material. And due to the ageing of its internal structure, the material suffered from different level of degeneration concerning other physical properties. Properties relying on late-wood resistance such as compression strength parallel to the grain and bending strength degenerated less heavily and maintained good uniformity while properties relying on early-wood resistance such as tensile strength paralleled to grain degenerated more and maintained good uniformity as well. On the other hand, properties relying on both late wood and early wood such as splitting strength and impact hardness had much poorer uniformity. This points out that the timing takes a great toll on physical properties of wood material.

Parameters	Old wood in Yingxian Wooden Pagoda	New wood	Old/new (%)
Compress strength parallel to the grain (kgf/cm^2)	467.7	576	81
Chordwise compress strength perpendicular to the grain (kgf/cm^2)	15.8	84	19
Radial compress strength perpendicular to the grain (kgf/cm^2)	20.6	46	45
Tensile strength parallel to the grain (kgf/cm^2)	651.7	1299	50
Bending strength perpendicular to the grain (kgf/cm^2)	964.7	1133	85
Chordwise shear strength parallel to the grain (kgf/cm^2)#	96.2	68	110
Radial shear strength parallel to the grain (kgf/cm^2)	89.13	85	105
Radial splitting strength (kgf/cm^2)	7.4	10.1	73
Chordwise impact hardness (kgf)	127.7	425.7	30
Radial impact hardness (kgf)	113.9	227.8	50
End-face impact hardness (kgf/cm^2)	433	377	115

Table 1. Comparison between old wood in Yingxian Wooden Pagoda and new wood in Ref. [15].

In 1994, Ni et al. ran chemical component analysis on the replaced columns from the main hall of Bei Yue Temple and Da Bei Lou building in Chang Ling, Hebei province, in northern China during renovation, see Ref. [16]. The two columns were, respectively, 900 and 200 years old and made of Chinese spruce (*Picea asperata*) and cypress (*C. funebris Endl.*). Samples in the experiments were taken from the intact part of the column. Results are seen in **Table 3**.

Parameters	Old wood in Jingqing Gate	New wood	Old/new (%)
Compress strength parallel to the grain (kgf/cm^2)	539	427	126
Chordwise compress strength perpendicular to the grain (kgf/cm^2)	42.6	49	87
Radial compress strength perpendicular to the grain (kgf/cm^2)	57.3	65	88
Tensile strength parallel to the grain (kgf/cm^2)	450	1070	42
Bending strength perpendicular to the grain (kgf/cm^2)	267	796	34
Chordwise shear strength parallel to the grain (kgf/cm^2)	108	73	148
Radial shear strength parallel to the grain (kgf/cm^2)	105	95	110
Radial splitting strength (kgf/cm^2)	12.1	12.2	99
Chordwise splitting strength (kgf/cm^2)	13.6	15.8	86
Chord plane hardness (kgf)	372	242	154
Radial plane hardness (kgf)	399	227	130
End-face impact hardness (kgf)	509	306	166

Table 2. Comparison between old wood in Jingqing Gate and new wood in Ref. [15].

It can be inferred from the data that various extract amounts from old wood showed different levels of increase while the amount of holocellulose and α-cellulose decreased, which showed that the main components of cell walls in old wood had degraded and had looser structure than newly lumbered ones. Cellulose is the main cause of high tensile strength parallel to the grain, and hemicellulose and lignin give the material elasticity and compression strength so the decrease of these three components microscopically explains the macrolevel mechanical properties degeneration.

3.2. Study on the decay pattern of physical properties, residual strength and longevity of wood material

Due to the special fact that ancient buildings are being reserved, old materials in studies are mostly the small components being replaced during renovation, which severely restrained the study on the strength of wooden structures. And the fact that the strength alters differently under different load conditions makes it even more difficult to study the decay of physical properties of old wood structures.

In 2006, Liu et al. did a study on the relevance between chemical components and bending strength and degree of decay on the old materials from the Wu Ying Palace in the Forbidden City, see Ref. [17]. The experiment samples were from the beam and made of larch. And the

Component	Tree species					
	Spruce			Cypress		
	Old	New	Old/new (%)	Old	New	Old/new (%)
Moisture content (%)	6.65	6.02	110.4	6.19	7.57	81.8
Ash content (%)	0.42	0.78	53.8	0.58	0.41	141.4
Cold water extract (%)	5.53	1.42	389.4	6.69	3.42	195.6
Hot water extract (%)	7.27	2.68	271.2	7.98	4.56	175.0
Phenethyl alcohol extract (%)	6.60	1.63	404.9	6.35	6.90	92.0
1% NaOH extract (%)	25.1	12.4	202.4	23.5	17.1	137.4
Pentosane (%)	11.5	11.6	99.1	16.6	10.7	155.1
Lignin (%)	30.0	28.4	106	33.1	32.4	102.1
Holocellulose (%)	58.6	66.2	88.5	56.6	64.9	87.2
α-Cellulose (%)	36.2	41.5	87.2	34.9	39.1	89.2

Table 3. Comparison of chemical composition between old wood and new wood in Ref. [16].

decay degrees were determined according to *GB/T 13942.2-92*, see Ref. [18]. The relationship between decay degree and cellulose and the content of 1% NaOH extracts can be seen in **Figure 4**. Due to the limited amount of old wood materials, chemical components analysis was carried out on healthy wood with a bending strength of 90, 100 and 110 MPa. The results of the relevance between bending strength and cellulose and the content of 1% NaOH extracts are shown in **Figure 5**. As the results showed, as the decay degree elevates, 1% NaOH extracts content evidently rises. Positive proportional relation can also be observed between 1% NaOH extracts and bending strength. The study showed that not only can the alkali extracts be used to determine the preliminary decay degree but they can also be used to determine visually healthy materials' physical properties.

To address the issue that in physical property experiments on ancient buildings, old materials are rare and the material qualities of new materials are different from those of the old ones, Xu et al. came up with the solution of accelerating the decay process via inoculation of fungus, see Ref. [19]. The process is to infect the wood with single fungus under suitable environment to accelerate the decay. This study provides the physical properties of wood from different decay degrees and the decay patterns of physical properties of wood from different decay degrees. And it offers a new way of thinking about the quantification of wood decay degree.

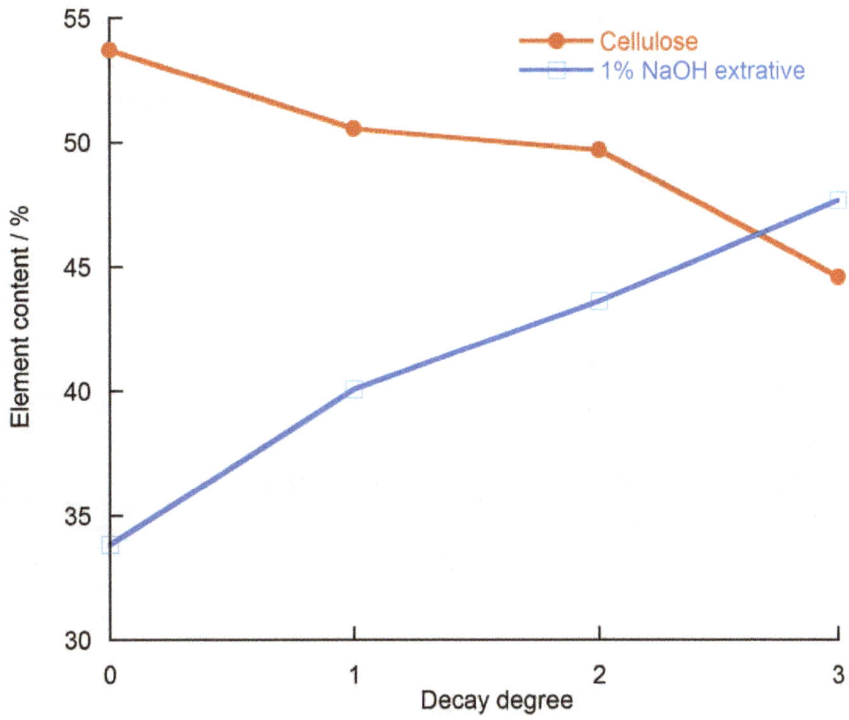

Figure 4. Variation in cellulose and 1% NaOH extract along decay of larch wood in Ref. [17].

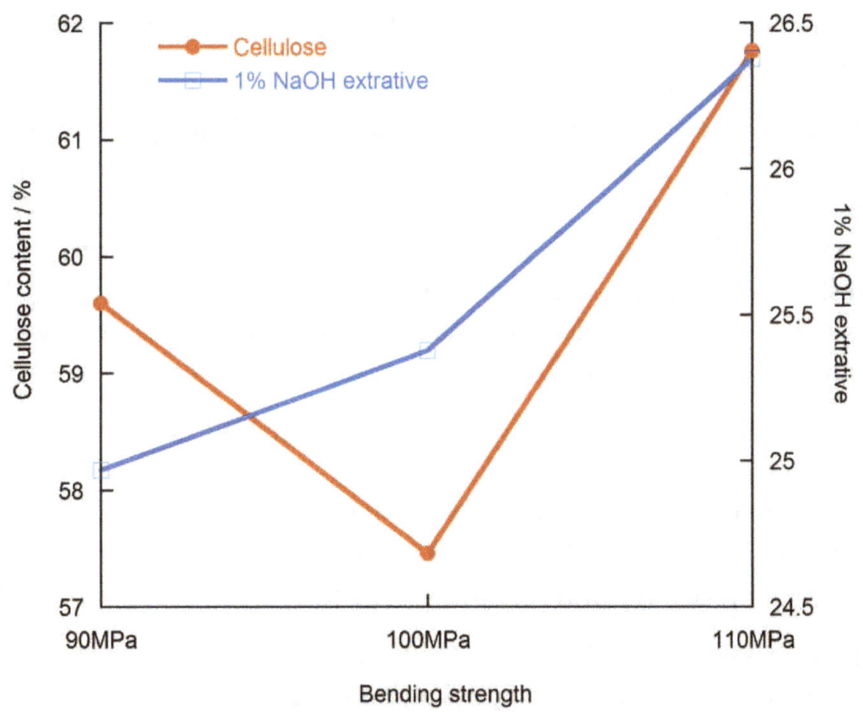

Figure 5. Variation in cellulose and 1% NaOH extract along bending strength of larch wood in Ref. [17].

3.3. Study on modern engineering materials' application in traditional wooden structure

In modern China, all the constructions of new palaces and temples involve reconstruction of pre-existing historic buildings or reference to classic elements. This requires similar construction method as well as high-quality materials. Among the new engineering wood materials, glued-laminated timber (glulam) shows all kinds of advantages such as natural wood texture, high quality of anti-corrosion, high usage and stable physical properties. In addition, glulam

(a) Main hall of Xiangji Temple in Hangzhou

(b) Main hall of the White Lotus Lore Temple in Shanghai

Figure 6. Typical applications of glulam used in Chinese traditional structure.

also has great plasticity and special expressive ability that can rival steel structure. Being well liked both worldwide and here in China, glulam has been applied to the construction of traditional structures, see Ref. [20].

Xiangji Temple is a historically famous temple in Hangzhou, Zhejiang province, in south-eastern China. First built in 1016 AD, the original building was destroyed in a fire and was rebuilt in 2010 AD. The major structures of the temple's bell tower, drum tower and the Kinnara hall were built in steel while the monastery, the guest house and the dorm rooms were in log structure and the rest were in glulam structure. Glulam combined with traditional roof made a column-free room possible. With the traditional multi-overhanging eaves, the main hall demonstrates a splendid momentum as well as openness and brightness, as seen in **Figure 6(a)**.

In the White Lotus Lore Temple in Shanghai, the Buddha hall has a glulam body. Different from the traditional temples, this temple represents its own era (as seen in **Figure 6(b)**). Built up high, with large overhanging eaves and mild-slope roof, it was totally built according to the proportion of buildings in Tang Dynasty. It is completely made of glulam and structured with grid structure to make a column-free indoor space.

The Yu Xi Temple Tower in Chao Mountain in Hangzhou, Zhejiang province, in south-eastern China is a pavilion-style tower with an octagonal section and a pillar in the centre. The tower has five floors with four additional floors hidden inside. It was made of glulam and the hidden floors were of structural reinforcing purpose just like those in the Yingxian Wooden

Pagoda. The Laojun tower in Qingcheng Mountain, Sichuan province, in south-western China is another wooden structure bu`ilding built on top of a mountain. After reconstruction, it is 28.05-m high, with a reinforced concrete base. The first and second floors are made of concrete and glulam and from the third through ninth floors are made of glulam, with Douglas fir.

Thanks to the variety of materials and connection types, traditional wooden structures are more frequently combined in modern construction. On the one hand, wood is combined with glass, concrete and steel, resulting in more flexibility in wooden structure design. On the other hand, wooden structure borrowed the steel system such as the grid structure in the White Lotus Lore Temple and the truss system in Kai Yuan Temple which extend the building scale. What's more, traditional mortise and tenon joints are combined with metal joints and adhesives. Through structural innovation and optimization, with advanced construction techniques, contemporary traditional wooden structures will be perfect in structural logic, creativity and details.

4. Anti-seismic performance study on traditional Chinese wooden structure

Wooden structures have an outstanding advantage over other forms of structures when it comes to anti-seismic performance. Traditional Chinese wooden buildings have a unique form of structure that allows them to withstand earthquakes with remarkable stability, hence the saying 'The building stands even though all its walls collapse'. One of the most significant features of traditional Chinese wooden structure is that it 'emphasizes structural members rather than joints', so the mechanical properties of joints take a huge toll on the performance of the whole building. This chapter shows studies on anti-seismic performance of key joints and whole buildings.

4.1. The anti-seismic structure and mechanics of ancient Chinese wooden structures

After analysing the damage that past earthquakes did on existing ancient wooden structures, experts found that ancient Chinese wooden structures have their unique features in design concepts, structural layout and building techniques. Special building techniques such as the floated joint between a column and the stylobate, the semi-rigid mortise and tenon joint between a beam and a column and the tilt columns and raised columns of the column frame along with the Queti, a kind of trimming joists at the end of a beam, the Dou-gong bracket and the 'grand roof', make classic wooden buildings distinct from modern reinforced concrete structures regarding anti-seismic performance.

With relatively high ratio of strength to weight, wooden material can maintain a certain level of resilience and ability to recover from deformation when external forces are applied. The mostly used joint formation between wooden components is semi-rigid mortise and tenon joint which not only improves the resilience of the whole structure but also effectively cancels the horizontal thrust and consumes a notable share of energy generated by the friction and rotation of mortises and tenons. Besides, classic Chinese wooden structure can also consume and absorb seismic energy through the auto-deformation of its load-bearing frame system.

Looking at the small components of the structure, it can be found that the connection between the column and the floor is often smooth and horizontal with no embedment or adhesion, which allows the upper section of the building to slide independently and stably as a whole during an earthquake without collapsing. A tilt column means to make the bottom of a column into a gentle slope, resulting in the top of it tilting slightly inward, the mortises and tenons above pressed together and the deadweight of the mortises and tenons providing the original bending moment of the joint. It can also act as an effective limitation on the movement of the upper beam frame. As the transitional layer between the column frame layer and the beam frame layer, the Dou-gong bracket layer is constructed of many small components interlaced, forming an upside-down triangle by using less and less components from top to bottom. It functions as a spring cushion and reduces the earthquake effect. And because of the transition and separation of the Dou-gong bracket layer, the roof and the beam frame as a whole can be analysed as a rigid entirety with slopes, as seen in **Figure 7**.

Before the 1990s, out of the purpose of reserving cultural heritage, studies on ancient buildings mostly highlighted their historical and artistic qualities rather than scientifically analysing their structures. In 1991, Wang T's analysis (see Ref. [21]) of the static load performance of the critical components, joints and the whole structure of ancient buildings marked the beginning of structural studies on ancient buildings. And studies on ancient Chinese structures have thrived so far. Focusing on the outstanding anti-seismic quality of classic wooden structures, Fan from the Harbin Institute of Technology, Yu and Xue from the Xi'an Jiaotong University, Zhao and Zhang from the Xi`an University of Architecture and Technology and Fang from the Tsinghua University conducted a large amount of experiments and theoretical analyses on the dynamic features, anti-seismic behaviours, destruction assessment and joint reinforcement. Li from the Taiyuan University of Technology and Zhou from the Peking University, respectively, conducted years of anti-seismic and reinforcing restoration experiments and studies on the Yingxian Wooden Pagoda and ancient buildings in the Forbidden City, as seen in Refs. [22–25], for example. Currently, the most commonly used methods of analysing the anti-seismic behaviours of classic wooden structures are static procedure, response spectrum analysis, dynamic-timing analysis and nonlinear static procedure. After years of studying, scholars at home and abroad have come up with methods of building analytical modules such as the semi-rigid calculation module of mortise and tenon joints, the combination module of beam units, the single degree of freedom (SDOF) system module and the mechanics module.

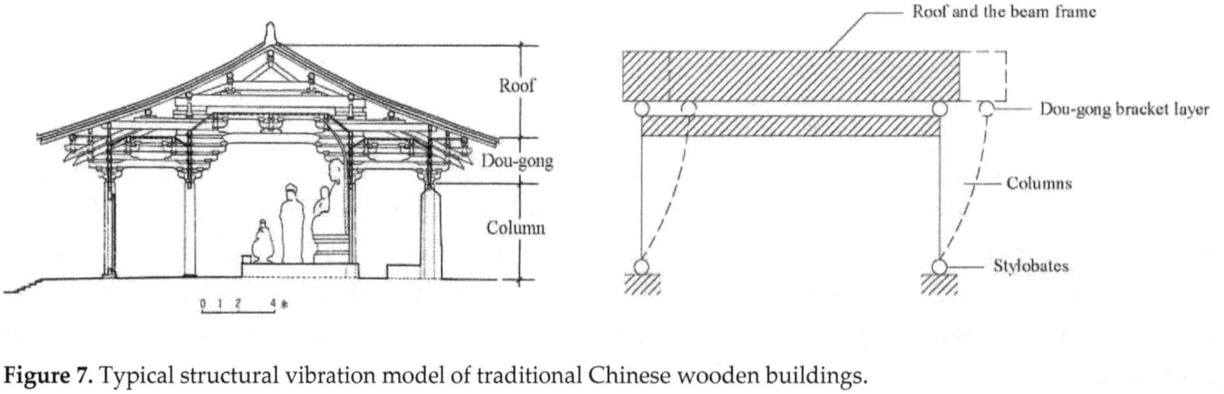

Figure 7. Typical structural vibration model of traditional Chinese wooden buildings.

4.2. The anti-seismic performance study on mortise and tenon joints and Dou-gong brackets

4.2.1. Mortise and tenon joints

Mortise and tenon joints are often used in classic buildings to join beams and columns. These joints can bear some lateral load and joint-bending moment and allow some rotation and relative slide between the beam and the column. These are all 'semi-rigid' features that this type of joints demonstrates, which can consume part of the energy and reduce structural reaction to earthquakes.

As to experiments, Gao et al., see Ref. [26], conducted lateral low-cyclic reversed-loading tests on three wooden structure models with Queti of the watchtower in Xi'an, Shaanxi province, in north-western China. They analysed the deformation features and destruction pattern of joints and found after calculation that the ductility coefficient changes within the range of 1.58–3.99. Xie et al., see Ref. [27], conducted the same experiments on dovetail joint models and discussed the effect on the anti-seismic performance of joints of vertical load, Queti, Pupai-fang components and the module size effect. As to calculation module, Wang simplified mortise and tenon joints as hinges and Queti as cantilevers with load focused on the tips in static calculation of wooden structures and double-checked the load-bearing capacity of components, as seen in Ref. [19]. Fang and Yu et al. built an FE model fit for ancient wooden buildings by defining 3D variable semi-rigid joints that reflect features of the Dou-gong bracket and mortise and tenon, based on studies on structural features of ancient wooden buildings, as seen in Ref. [28] and **Figure 8**. The module was first used in the calculation of the unequal settlement of the base of the watchtower in Xi'an and then used in the mechanics performance analysis of ancient buildings such as the drum tower in Xi'an, the Baoguo Temple in Ningbo and Zhejiang province, and performed quite a good job. Feng and Zhang et al. (see Ref. [29]) combined the shaking table test on the column frame unit model formed by four columns in the palace hall structure in building standards of Song Dynasty and low-cyclic reversed-loading tests on the model and numerical simulation, conducted lateral vibration analysis and random destruction theoretical analysis on the Dou-gong brackets and studied the features of semi-rigid mortise and tenon joints. And came up with the rigidity formula for mortise and tenon joints and the equivalent viscous damping coefficient.

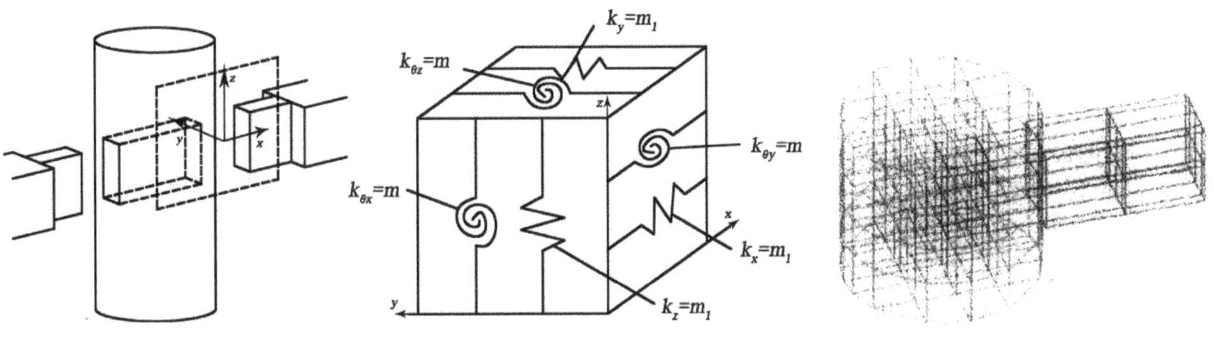

Figure 8. Semi-rigid joint model and finite element modelling of mortise and tenon joint in Ref. [29].

4.2.2. Dou-gong brackets

Dou-gong bracket, a special connection component between the column and the beam, plays a pivotal role in both structural force transmission and decorative function. Composed by many cantilever joists (named as Gong) staked one on top of another on crossed direction and

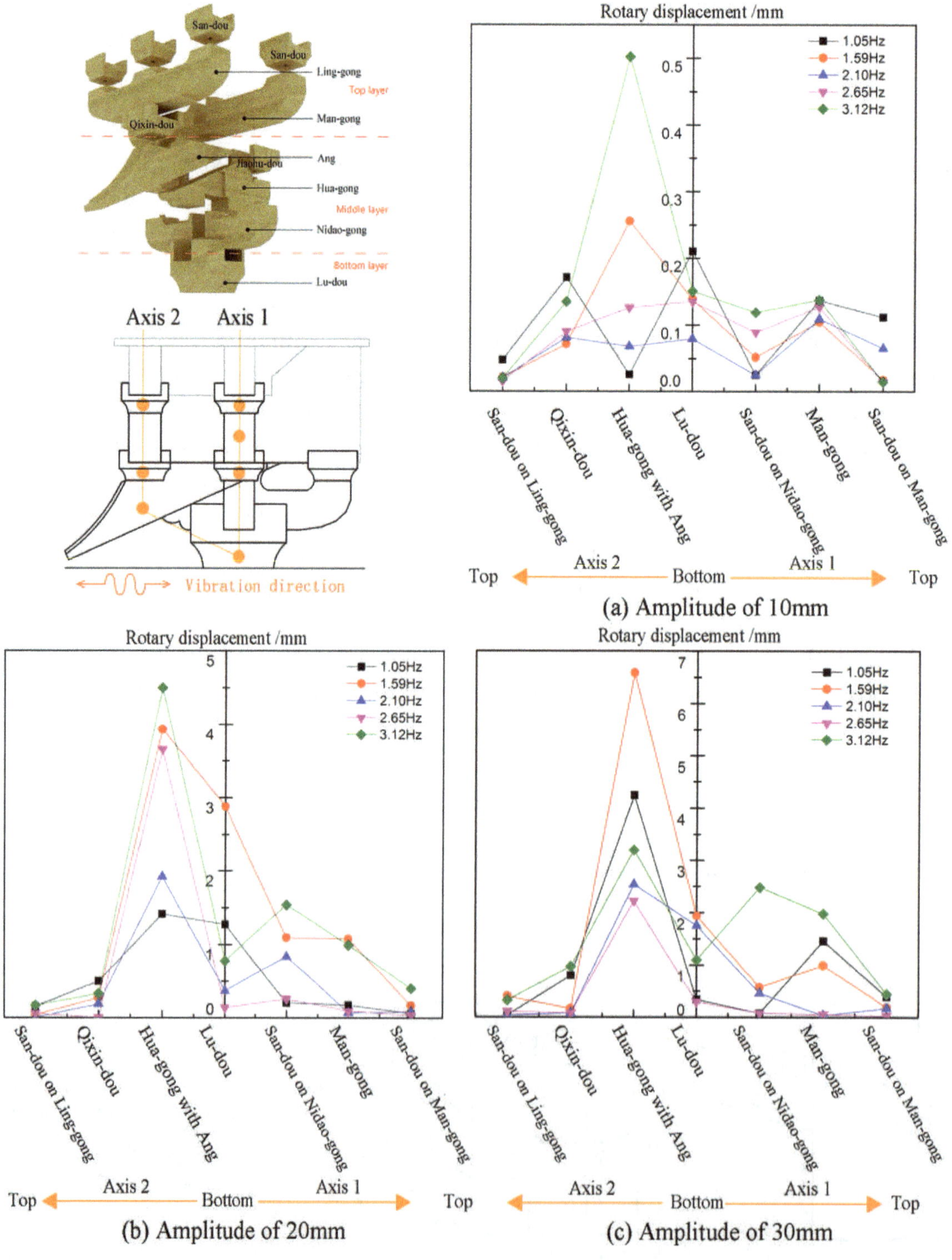

Figure 9. The rotary displacement of the Dou-gong components in shaking table tests along two axes in Ref. [33].

connected by Dou members, Dou-gong bracket as a whole could just be regarded as a beam pad. This special structure forms as an inverted fixed-hinged support which has compression deflection and rotary movement on vertical plan as well as slip movement on horizontal plan. In respect of structural performance, because of the overhanging on two directions, Dou-gong bracket shortens the span and enhances the load-carrying capacity of upper beams; meanwhile, it helps to adjust the depth of eaves, making it more graceful and harmonious. On the other hand, instead of sticking together, all the Dou and Gong elements are connected by mortise and tenon joints. With the addition of its unique shape with overlapping cantilevers, Dou-gong bracket becomes a ductile connection to dissipate energy between column and beam, especially under lateral forces such as earthquakes.

Chinese scholars mainly conducted studies on two types of Dou-gong brackets: the Song style and the Qing style. During the restoration project of the tower of the east city gate in Xi'an in 1996, with the help of the repair team, Yu et al. conducted and performed static and dynamic experiments on the two types of Dou-gong brackets, see Ref. [30]. Gao et al. performed six 1:3.52 models of the bottom two pieces of Song style 8 layers Dou-gong brackets with second-class material according to the regulations in building standards and came up with load-displacement calculation module, mass-spring-damper model and lateral force-displacement-restoring force model under vertical load via vertical monotonic-loading tests and lateral low-cyclic reversed-loading tests. They also calculated the vertical seismic transmission coefficient and the lateral energy consumption, which showed great two directional anti-seismic performance of the Dou-gong brackets, see Ref. [31]. Sui et al. drew the conclusion of a restoring force model that reflected the restoring features and stiffness variation regulation of Dou-gong through low-cyclic reversed-loading tests on singular-layer, double-layer and quadruple-layer Dou-gong models, see Ref. [32]. As to other transitional types of Dou-gong brackets in the transition periods, our team of the Nanjing Forestry University conducted shaking table tests on full-scale models made of Douglas fir and China fir based on the Dou-gong brackets in the Ming Dynasty Tian Wang Palace of Bao Sheng Temple in Luzhi and analysed factors such as the between-layer displacement reaction features, the contribution ratio of the rotation and slide deformation of the components and the structurally weak-part assessment as seen in Ref. [33] and **Figure 9**.

As to numerical simulation, Wei, see Ref. [34], studied the non-linear variation patterns of the connection rigidity, calculated the ductility coefficient and the equivalent viscous-damping coefficient and compared results between the axial compression tests and the low-cyclic reversed-loading tests via ANSYS simulation based on the operating mechanism, destruction form and anti-seismic performance of Dou-gong. Du studied the Yingxian Wooden Pagoda, built the simplified models of the rigid connection and hinges in a Dou-gong bracket using the dynamic equivalent features method and calculated the range of the dynamic features though the two simplified models and then applied them to the calculation of the whole tower, see Ref. [35].

5. Conservation and reinforcement techniques of historic wooden buildings in China

5.1. The principles of historic architecture restoration

Since 1920s, China began to attach importance to the conservation of historic wooden buildings. In 1928, the *Central Commission for the Preservation of Antiquities* was established. In 1929,

Zhu et al. founded the *Society of the Study of Chinese Architecture*. In 1930, the government issued the *Regulation of Antiquities Conservation*, which symbolized the start of legal management of antiquities. The *Law of the People's Republic of China on the Protection of Cultural Relics* was promulgated in 1982, including the protection of antiquities in the law, which symbolized the standardization and internationalization of historic wooden buildings protection. At present, the protection of historic wooden buildings in China mainly follows the *International Charter for the Conservation and Restoration of Monuments and Sites, Law of the People's Republic of China on the Protection of Cultural Relics* and the standard *GB 50165-92*. Maintenance and reinforcement construction comprises regular maintenance project, major project of historic preservation and maintenance, partial restoration project, relocation project and emergency project, and abides by the principle of maintaining the buildings' original state, including (1) original architectural form such as plane layout, modelling, construction characteristic, artistic style and so on, (2) original building structure, (3) original building materials and (4) processing technology (see Ref. [18]).

5.2. Traditional reinforcement techniques of historic wooden buildings

5.2.1. Common damages of historic wooden structures

The common damages of historic wooden structures include (1) component deformation under compression: column yielding such as splitting of column tips under compression, decay on the bottom of columns due to long-term exposure to humidity and splitting along the grain on the body of columns; bending and splitting of girders and square beams between column and Dou-gong brackets; breaking and splitting of subcomponents of Dou-gong brackets. (2) Components under tension: square beams through the columns on the top or the bottom usually get adrift or break off at the mortise and tenon joints. (3) Components under shear strength: the force analyses at the mortise and tenon joints are more complex and the joints tend to detach under long-term shearing action, as seen in Ref. [15].

5.2.2. Traditional reinforcement techniques of ancient wooden buildings

Regarding the whole-beam frame of wooden structures, reinforcement methods include major repair of the structure (disassembling the wooden frame completely of partially, repairing or replacing the damaged components and reassembling it while reinforcing the structure), restoration with external support (adding external support and restoring the tilted, twisted or detached components while reinforcing the structure without disassembling the frame) and overall reinforcement (direct reinforcement of the whole structure of projects with minimum structural deformation).

Reinforcement methods are as follows (**Figure 10**):

1. Partial or complete replacement: (a) patching and reinforcing. It can be patched up with wooden powder and waterproof adhesives when the splits or corrosion of beams and columns are slight. (b) Reattachment of columns: Replace the rotten part of the column with new materials when the rotten part takes up more than a quarter of the height. The spot where reattachment is conducted is often reinforced with a semi-tenon and an iron hoop. (c) When the damage depth of the beam at both sides takes up more than a third of the height, it is appropriate to use the clamp connection method but when the depth takes up more than three-fifths, a replacement of the beam head is necessary.

2. Mechanical reinforcement: (a) Ironware reinforcement: Using flat iron to reinforce beams and columns or to connect joints between beams and columns. This way, the flat iron can improve the mechanical properties of components by bearing part of the tensile, compressing, bending and shearing forces. (b) When the deflection of beams and square pillars transcends normal limitation or the load-bearing capacity is insufficient or splits are found, it is appropriate to use tensile bars to form now load-bearing components.

3. Chemical reinforcement: Since the 1970s, unsaturated polyester resin filling was widely used in historic building restoration. Via the filling, soaking, patching or painting of chemicals, not only can the strength of the damaged wood be improved but the stability and antirot capacity can also be enhanced. In the restoration of the main palace of Nanchan Temple in Wutai, Shanxi province, 782 AD in 1974, epoxy resin was filled in the splits of two main beams and iron hoops were fixed on the beams. In the 1975 restoration of the main palace of Baoguo Temple in Ningbo, Zhejiang province, 1013 AD, without disassembling the wooden frame, the termite-ridden columns were filled with chemicals and wrapped in fibre-reinforced plastic (FRP). According to calculation, despite the larger expenses of chemical fillings, at least 30% of the budget was saved because disassembling the whole frame was avoided, see Ref. [36].

There are some disadvantages while using the above traditional reinforcement techniques. The ironware is usually applied inside the components and easily corroded so the appearance of the structure may be affected. The antirot chemicals greatly harm the health of the management staff. When using the reattaching method or the tensile bars, the original appearance of the building is inevitably damaged.

(1) Reattachment of columns and Mechanical reinforcement (2) Chemical reinforcement (3) CFRP sheet reinforcement

Figure 10. Reinforcement techniques of ancient wooden buildings.

5.3. The study on and application of FRP reinforcement techniques in historic wooden structures

Fibre-reinforced polymers (FRP) have advantages such as large tensile capacity and are light weight. It can also endure erosion, heat and freezing. Besides, it is highly plastic, easy to apply and inexpensive. As a result, it is widely used in the restoration of reinforced concrete and brick structures. The study on FRP started in the 1990s and has matured in the theory of reinforced concrete restoration. In the field of wooden structure reinforcement, FRP utilization especially carbon fibre-reinforced polymers (CFRP) and glass fibre-reinforced polymers (GFRP) utilization are becoming a heated study topic. Based on relevant studies and projects, FRP reinforcement can reduce several variation coefficients and strength indices of mechanical properties of wooden components. Variation coefficients after reinforcement are limited by 15%. FRP reinforcement can also improve the bearing capacity and reduce the long-term creep of wooden components, optimize the cross-section size and enhance the fire-resisting and antirot capacity.

FRP reinforcement can improve the anti-seismic capacity of mortise-tenon joints. Zhou et al. from The Palace Museum made a special frame model with four beams and four columns at the ratio of 1:8 using the Korean Pine and tenon-mortise connections based on the actual size of the partial frame of the Hall of Great Harmony in the Imperial Palace and conducted exploratory experiments using CFRP fabric. Through low-cyclic reversed-loading tests, load-displacement hysteretic curves were drawn and skeleton curves, energy dissipation capability and stiffness degradation of the structure were analysed. Results show that although the structural energy dissipation capacity decreased slightly after the model is strengthened by CFRP sheets on tenon-mortise location, the cross-section size of the tenon that will pull out of the joint reduces. Its lateral stiffness and load-bearing capacity both improve and with slight stiffness degradation, the frame still has good deformation capacity. Besides, the team also considered to conduct performance comparison among mortise-tenon joint models using nails, iron hoops and CFRP reinforcement. The results showed good deformation capacity preservation in all three scenarios. CFRP fabric excels in joint-bearing capacity and energy dissipation capability but showed the most stiffness degradation. So, it is recommended to use CFRP in reinforcement of medium to small wooden structures (see Ref. [37]).

Huang deducted the calculation of FRP anti-sheering reinforcement using basic materials mechanics formulas. Yang analysed the influence of CFRP and GFRP on the bend-resistant capacity and came up with the formula of ultimate bearing capacity for the analysing method based on failure strain. A team from the Xi'an University of Architecture and Technology did much work on the analysis of and tests on RFP reinforcement in historic building component units and structural joints. Case in point, Xie (see Ref. [38]) tested the bend-resistant capacity of square beams with CFRP reinforcement and compressive strength of cylinder columns and established the calculation of shear strength of beams and compressive strength of columns in different damage forms. Besides, he built scale models of the column frame according to the classic ancient architecture rules in Song Dynasty, 960–1270 AD and conducted low-cyclic reversed-loading tests on the original structure, CFRP sheets-reinforced structure and flat steel-reinforced structure, based on which the restoring force model of wooden structure was established. Hang et al. (see Ref. [39]) analysed the load-bearing performance of damaged joints reinforced with CFRP based on the above-mentioned-reinforcing approach and damage form of the joints, and came up with the calculation of the bend-resistant strength of the joints based on relevant experiments and calculation presumptions.

In the field of construction application, CFRP wrap combined with traditional wooden structure reinforcement methods has already been used on historic buildings such as the Tiananmen Gate tower and explored in a practical manner. In the emergency restoration of Yingxian Wooden Pagoda, in Fogong Temple, Shanxi province, 1056 AD, experts recommended FRP materials to maintain its original appearance. Combined with the characteristics of Chinese wooden structures, FRP reinforcement's performance in fire-resistant capacity and structural assessment under long-term load awaits further study.

6. Conclusions

Chinese traditional wooden architecture, well known as a unique and independent system of the architecture world, has formed its typical structural styles and construction technology after over 7000 years' development. Besides the research on and conservation and reinforcement status of many precious historical architectural heritages, this chapter is also focused on the analysis of structural features, anti-seismic behaviour and utilization of new materials in traditional wooden architecture based on a large number of studies in recent years.

To better protect ancient wooden structure, many researchers have carried out a large amount of physical properties experiments on wood materials from historic buildings, and a method for predicting the degeneration pattern of physical properties, residual strength and longevity of wood material through studying wood decay has been proposed. In addition to traditional reinforcement techniques such as mechanical reinforcement and partial or complete replacement, new reinforcement materials and techniques have already been explored, among which FRP is becoming a heated academic topic.

The superior seismic performance of Chinese traditional wooden architecture, owing to many unique characteristics of the structural design and constructional technique, has generated a great deal of interest at home and abroad. The objects and models of anti-seismic behaviour study also show characteristics of miniaturization and diversification. In addition to researches on historical wooden buildings, attempts of using new engineering wood products in the modern wooden architectures of traditional style are becoming a great upsurge nowadays.

At present, researches of Chinese traditional wooden structures have made some headway, yet there are also some remaining issues. First, material performance and structural behaviour study of historical wooden buildings are often based on specific emergent repairment and strengthening projects of historical buildings, which somewhat limits the systematization and universality of the researches. Second, attentions have been focused on historic and artistic aspects for a long time, and limited number of fundamental studies on the structural performance of Chinese traditional wooden structure and its typical connections types can be found. What's more, combine the excellent features of traditional construction technologies with modern materials, techniques, and then inherit and improve them, there is still much work for us to do.

Acknowledgements

This work was supported by the Special Fund of Top-Notch Academic Programs Project of Jiangsu Higher Education Institutions (TAPP) and National 'Twelfth Five-Year' Plan for Science & Technology Support (2015BAD14B0503).

Author details

Ze-li Que[1]*, Zhe-rui Li[1], Xiao-lan Zhang[1], Zi-ye Yuan[2] and Biao Pan[1]

*Address all correspondence to: zelique@163.com

1 College of Materials Science and Engineering, Nanjing Forestry University, Nanjing, China

2 School of Foreign Studies, Nanjing Forestry University, Nanjing, China

References

[1] Hao C.R. Perspective of wood buildings in China view from the development of Chinese and western wood buildings [thesis]. Beijing: Tsinghua University; 2004. 106 p (in Chinese).

[2] Pan G.X, editor. A history of Chinese architecture. 7th ed. Beijing: Architecture & Building Press; 2015. 555 p (in Chinese).

[3] Liu D.Z. History of ancient Chinese architecture. 2nd ed. Beijing: China Architecture & Building Press; 1984. 432 p (in Chinese).

[4] Ma B.J. The construction technology of China ancient wooden buildings. 2nd ed. Beijing: Science Press; 2003. 355 p (in Chinese).

[5] Chen M.D. Yingxian Wooden Pagoda. 1st ed. Beijing: Cultural Relics Press; 1966. 246 p (in Chinese).

[6] Li T.Y, Qin H.M. Structural analysis and repair of Yingxian Wooden Tower. Engineering Mechanics. 2005; **22**: 199–212 (in Chinese).

[7] Du L.M, Li H.W, Xue F, Qin D.Q. The study of seismic behaviors of Yingxian Wooden Pagoda. China Civil Engineering Journal. 2010; (43): 363–370 (in Chinese).

[8] Fan C.M, Wang L.A, Pan J.L. Measures to control seasoning checks of wood used for the repair of Yingxian Wooden Pagoda, Shanxi Province. Journal of Beijing Forestry University. 2006; **2**(1):98–102 (in Chinese).

[9] Chen Z.Y, Zhu E.C, Pan J.L. Lateral structural performance of Yingxian Wooden Pagoda based on refined FE models. Journal of Building Structures. 2013; **34**(9): 150–158 (in Chinese).

[10] Yu Z.M. Structure chart analysis of Yingxian Wooden Pagoda [dissertation]. 2014 (in Chinese).

[11] Hou W.D. Disputes and debates: conservation philosophy of the Yingxian Wooden Pagoda. World Architecture. 2014; (12): 38–41. DOI: Doi: 10.16414/j.wa.2014.12.005

[12] Chai J.Z. Architectural form analysis of the East Palace of Foguang temple. Mt Wutai Researches. 1986; (1): 17–20 (in Chinese).

[13] Zhang R. Survey of the east palace of Foguang Temple. Traditional Chinese Architecture and Gardens. 2010; (3): 29–39 (in Chinese).

[14] National Technical Committee 41 on Timber of Standardization Administrator of China. GB 50165-92 Technical code for maintenance and strengthening of ancient timber buildings. Beijing: China Building Industry Press; 1993. 90 p.

[15] Chen G.Y. Study on material properties change of old wood of historic buildings and the effect on building deformation. Traditional Chinese Architecture Gardens. 2003; (3): 49–52 (in Chinese).

[16] Ni S.Z., Li Y.Z. Research on tree species of old wood in traditional Chinese wooden architecture and analysis of material properties. Sichuan Building Science. 1994; (1): 11–14 (in Chinese).

[17] Liu W.B. Study on relationship of molder condition between variations of chemical ingredients and bending strength of ancient wood structure in the Imperial Palace [dissertation]. 2006 (in Chinese).

[18] National Technical Committee 41 on Timber of Standardization Administrator of China. GB/T 13942.2-92 Method for field test of natural durability of wood. Beijing: China Standard Press; 1993.

[19] Xu M.G, Qiu H.X. A new way of study on ageing of traditional Chinese architecture. Earthquake resistant engineering and retrofitting. 2009; **31**(2): 96–98. DOI: 1002-8412(2009)02-0096-03

[20] Ren L.Z, Liu J. The application of glued laminated timber in the design of contemporary Buddhist architecture: rehabilitation of Hangzhou Xiangji Temple. New Architecture. 2012: 86–89. DOI: 1000-3959 (2012) 05-0086-04

[21] Wang T. Preliminary exploration of static force of ancient Chinese architecture. Beijing: Cultural Relics Publishing House; 1992. 187 p (in Chinese).

[22] Yu M.H., Oda Y., Fang D.P., Zhao J.H., Zhang D.L., Zhu R.X., Che A.L. Advances in structural mechanics of Chinese ancient architectures. Advances in Mechanics. 2006; **36**(1): 43–64 (in Chinese).

[23] Chen Z.Y., Zhu E.C., Pan J.L. A review on structural mechanics of Chinese ancient wood structures. Advances in Mechanics. 2012; **42**(5): 644–654 (in Chinese).

[24] Xue J.Y., Zhang P.C., Zhao H.T. Study on the aseismic mechanism of historic timber structural building. J.Xi'an University of Architecture and Technology 2000; 32(1): 8–11 (in Chinese).

[25] Zhou Q., Yan W.M., Guan H.Z, Ji J.B. Research advances in Tou-Kung's mechanical performances of Chinese ancient wood buildings. Journal of Water Resources and Architectural Engineering. 2014; **12**(4): 18–26 (in Chinese). DOI: 10.3969/j.issn.1672-1144.2014.04.004

[26] Gao D.F., Deng H.X., Liu J., Li F., Yang Y. Pseudostatic experimental study on mortise and tenon joints of timber structures of Chinese Ming and Qing Dynasties. World Earthquake Engineering. 2014; 30: 9–16 (in Chinese).

[27] Xie Q.F., Du Bin, Xiang W., Zheng P.J., Cui Y.Z., Zhang F.L. Experimental study on seismic behavior and size effect of dovetail mortise-tenon joints of ancient timber buildings. Journal of Building Structures. 2015; 36: 112–120 (in Chinese).

[28] Fang D.P., Yu M.H., Yutaka M., Shoji I., Hediaki D. Numerical analysis on structural characteristics of ancient timber architecture. Engineering Mechanics. 2001; 18: 137–144.

[29] Feng J.L., Zhang H.Y., Wang H., Zhou H.D. Shaking table tests and analysis of Dougong layer of Chinese ancient structure. Sichuan Architecture. 2009; 29(4): 132–133.

[30] Zhao J.H., Yu M.H., Yang S.Y., Sun J.J. An experimental study for the dynamic characteristics of 'Dougong'—one of the wooden structure parts in ancient architecture of China. Journal of Experimental Mechanics. 1999; 14(1): 106–112 (in Chinese).

[31] Gao D.F., Zhao H.T., Xue J.Y. A seismic characteristics of bucket arch and mortise-tenon joint of ancient Chinese timber buildings experimental research. Journal of Natural Disasters. 2008; 17(2): 58–64. DOI: 10.13577/j.jnd.2008.0211

[32] Sui Y., Zhao H.T., Xue J.Y., Xie Q.F., Liu Y. Experimental study on lateral stiffness of Dougong layer in Chinese historic buildings. Engineering Mechanics. 2010; 27(3): 74–78.

[33] Que Z.L., Li Z.R., Zhang B.B., Hou T.Y., Pan B. Experimental study on shaking table tests of Dougong in Tianwang Hall, Luzhi, Ming Dynasty. Journal of Civil, Architectural & Environmental Engineering. 2015; 37(3): 26–34. DOI: 10.11835/j.issn.1674-4764.2015.03.004

[34] Wei G.A. Mechanical behavior and ANSYS analysis of Dougong in Chinese ancient timber building [thesis]. Xi'an: Xi'an University of Architecture and Technology; 2007 (in Chinese).

[35] Du L.M., Li H.W., Xue F., Qin D.Q. The study of seismic behaviors of Yingxian wooden pagoda. China Civil Engineering Journal. 2010; 43: 364–370 (in Chinese).

[36] Chun Q., Yu M.Z., Pan J.W. Research on damage characteristic and structural performance of the main hall of Baoguo Temple in Ningbo. Sciences of Conservation and Archaeology. 2013; 25(2): 45–51 (in Chinese).

[37] Zhou Q., Yan W.M., Ji J.B. Aseismic behaviors of tenon-mortise joints in wooden frame in Chinese ancient building strengthened by three materials. Journal of Building Materials. 2013; 16(4): 649–656 (in Chinese).

[38] Xie Q.F. Experimental study and theoretical analysis on strengthening for Chinese ancient timber buildings. College of Civil Engineering, Xi'an University of Architecture & Technology, 2007.

[39] Zhang F.L. Research on strengthening and its performance of Chinese ancient timber structures [dissertation]. Xi'an: Xi'an University of Architecture & Technology; 2013. pp. 194 (in Chinese).

7

Grading of Low-Quality Wood for Use in Structural Elements

Trulli Nicoletta, Monica Valdés,
Barbara De Nicolo and Massimo Fragiacomo

Abstract

Timber is a sustainable resource, environmentally friendly and aesthetically pleasing. Using locally grown timber as building material leads to economic, social and environmental benefits. Being an organic material, timber is not homogeneous; hence, it is crucial to predict the base material quality. International codes require the use of wood previously graded according to the current regulations in order to verify its reliability when used as structural material. An exhaustive analysis of the state of art of different methodologies and code requirements for structural timber grading is presented herein. Structural timber grading methods and their applicability to low-strength timber is analysed and discussed with reference to Maritime Pine locally grown in Sardinia (Italy). Several physical and morphological parameters such as density, the presence of knots, clusters of knots, grain deviation, warping, annual ring width and moisture content had to be measured. Moreover, mechanical parameters (tensile strength and modulus of elasticity in tension) were measured and analysed in order to identify the strength class of Sardinian Maritime Pine. The operational issues related to the application of the different methodologies and code requirements for structural grading of low-quality wood are also discussed and analysed.

Keywords: timber structures, visual and machine grading, low-quality wood

1. Introduction

In the last decades, timber is increasingly being used as building material as it represents a sustainable resource and is environmentally friendly and aesthetically pleasing when used both in new buildings and renovation. Using timber as building material leads to environmental benefits in terms of CO_2 emissions. During the growth, the trees absorb CO_2 by storing carbon and releasing oxygen in the atmosphere. When a tree is cut and processed into a building material, it delays the time when the carbon captured during the photosynthesis will be released back into the atmosphere. According to scientific studies and as shown in the Sixth Environmental Action Programme of the European Union, a cubic metre of wood used as construction material is equivalent to 1 ton of CO_2 that is stored instead of being released into the atmosphere [1–3].

Timber is also characterized by excellent properties such as lightness, low density, high strength-to-weight ratio, etc. These properties lead to the possibility to realize lightweight structures having excellent earthquake resistance, reduced cost of foundations, and the ease of transport and erection.

Due to the aforementioned advantages nowadays, a significant increase in the volume of timber is used in building structures even in countries where there was weak tradition in construction of wooden structures (e.g. Italy, Spain, France, etc.). This growth has also been made possible by the availability on the market of a wide range of wooden products such as cross-laminated timber (CLT) and glue-laminated timber (GLT) elements. However, most of the timber used in these countries is imported from abroad. Using locally grown timber as building material would lead to economic, social and environmental benefits.

Due to its organic nature, timber is not homogeneous; hence, it becomes of utmost importance to predict the base material quality and properties. The properties strongly depend on the growth condition and vary among different wood species [4]. International codes require the use of wood previously graded according to the current regulations in order to verify its reliability when used as structural material. Moreover, in Europe, structural timber shall be CE marked according to the European Construction Products Regulation (CPR) [5]. For these reasons, in the last 10 years, extensive researches have been carried out on locally grown timber species aiming at assessing the opportunity of a safe and economic use of these species as structural material.

In Europe, the procedure for grading structural timber is defined by EN 14081-1 [6]. There are two systems for timber grading: machine and visual grading. Both the systems define grades to which characteristic values of strength, stiffness and density can be allocated according to EN 338 [7]. Characteristic values of strength, stiffness and density can be defined and measured according to EN 384 [8]. The two grading systems differ in (i) the property measured to define the grading criteria and (ii) the normative requirements.

Visual and machine grading are based on defining visually and non-destructive parameters which are related to the three determining properties (density, stiffness, strength) based on relationships derived by means of destructive testing.

Visual strength grading requires the non-destructive assessment of each piece of timber in order to define grading rules by means of visual features such as knots, rings width, slope of grain, warping, etc. Grading rules specify limits for all these features in order to assign each piece of timber into a grade. Then, based on the result of the destructive test, the grades are assigned to strength classes according to EN 338 [7]. Visual grading can be applied and implemented simply and without special measuring equipment.

Machine strength grading is a process where a piece of timber is non-destructively sorted by a machine into grades by means of powerful predictors of the quality of the base material which are closely related to one or more of the grade determining properties. Machine grading can be applied quicker and with less risk of human error than visual grading.

2. Visual strength grading

In Europe, visual strength grading is performed according to EN 14081-1 [6] where minimum requirements for national visual stress grading standards are defined. Annex A sets the limitations for strength-reducing (knots, slope of grain, density and rate of growth, fissures), geometrical (wane, warp), biological (fungal and insect damage) and other (reaction wood) characteristics. The testing methods for determining the mechanical properties of sawn timber are specified in EN 408 [9]. The common strength class system is defined by EN 338 [7], while EN 1912 [10] sets up the assignment of species and visual grades derived from national standards to strength classes.

In general, each European country has developed its own grading rules to define the methods for measuring properties and their limits.

For coniferous sawn timber, as an example, the Italian Standard UNI 11035-2 [11] sets three grades (S1, S2 and S3), the Spanish Standard UNE 56544 [12] defines two grades (ME1 and ME2), while the German Standard DIN 4074-1 [13] establishes three grades (S13, S10, and S7).

2.1. Strength-reducing parameters

2.1.1. Knots

Knots are caused by a branch embedded in the log. Knots are classified according to their shape, size and position in sawn timber [14]. Knots size is one of the main parameters for visual grading of sawn timber because it tends to cause a downgrading of the sawn timber due to its effect on warping and strength. Several studies have been carried out aiming at verifying the knots effects on the mechanical properties of sawn timber. For Portuguese Maritime Pine timber, the extensive presence of knots in the boards caused a rejection of 50% of sawn timber during the visual strength grading procedure and 44% of downgrading in visual strength grades [15]. In softwood, an increase in knots size from 25 to 75 mm can cause a decreasing in bending strength up to 50% [16]. Moreover, as reported by Olsson et al. [17] during fracture testing of 1000 pieces of timber, more than 90% of the failures were caused by knots.

The criteria adopted in national standards for defining size of knots and their limitations are different. For example, the Italian Standard [11] defines the A and A_g parameters for knot and knot cluster, respectively. A signifies the ratio of the minimum knot diameter to the width of the cross section where the knot appears, while A_g denotes the ratio of the sum of the minimum diameters of the knots, comprised in a stretch of 150 mm, to the width of the cross section where the knot appears. According to the Spanish Standard [12], the limitation of the knots size depends on the type of knot (face, edge or margin knots): the knot diameter is defined as the distance between two straight lines tangent to the knot and parallel to the axis of the section. Additionally, the Spanish Standard [12] takes the knot cluster into consideration. The German Standard [13] defines a knot coefficient A in a simple way as the ratio of appropriate knot dimensions depending on the knot location to one of the cross-sectional dimensions. In the Polish Standard, as reported by Krzosek [18], the knot coefficient is given by a combination of two coefficients, tKAR and mKAR. tKAR coefficient is given by the ratio of the knot area in the weakest cross section to the cross-sectional area of the entire cross section. The parameter mKAR is the ratio of the cross-sectional area of the knots located in worse margin, namely closer to the corner of the cross section, to the area of the cross section. **Figure 1** shows the knot measurements in accordance with the Spanish, German and Polish Standards.

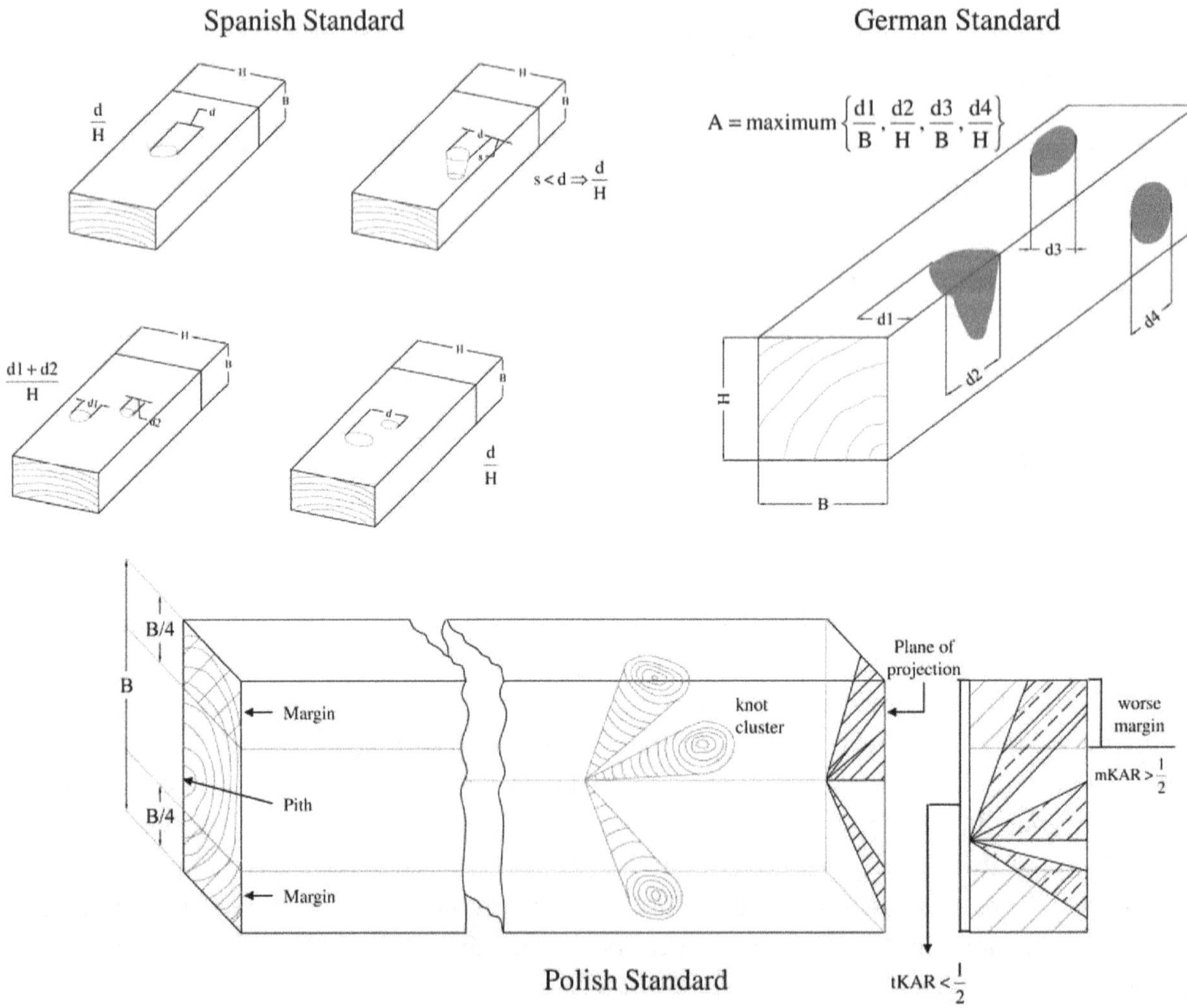

Figure 1. Knot coefficient according to different national standards.

Comparisons of the results of visual strength grading rules applying the different national standards have been developed by researchers. Adell Almazán et al. [19] compared the results of Scots Pine visual strength grading according to the Spanish and German Standards and found a large difference. They stated that the most critical parameter was related to knot: 40% of the sawn timber pieces was rejected using the Spanish Standard, while only 5% of the sample was rejected following the indication of the German one. Krzosek [18] found irrelevant difference related to knot when applying the Polish and the German Standards on visual strength grading of *Pinus sylvestris* sawn timber. Stapel et al. [20] compared the results of visual strength grading of softwood sawn timber according to the German, Swiss, British, Danish and French Standards and pointed out several differences in the results caused by different rules of measuring knots and to an unequal number of visual grades in the standards.

2.1.2. Slope of grain

Slope of grain is a deviation of wood fibres from a line parallel to an edge of sawn timber. Slope of grain is expressed by the ratio between the deviation of grain length in millimetres (x) and the length over which the measurement is taken, in millimetres (y) as shown in **Figure 2.**

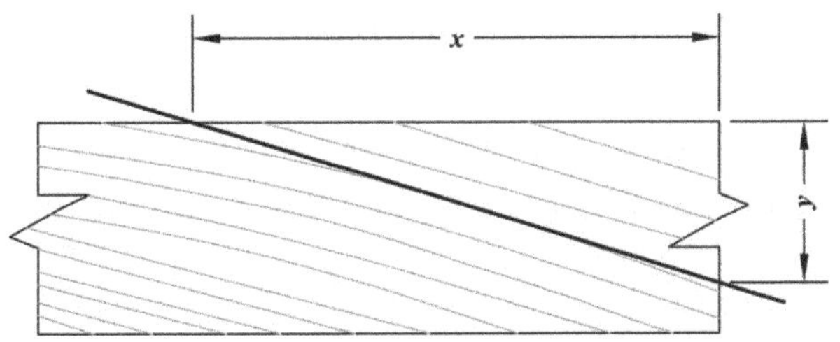

Figure 2. Slope of grain measurement.

Slope of grain is markedly depending upon wood species and is generally caused by two sources:

- slight bend of the tree: wood cells are arranged at a slight angle with respect to the axis of the stem during the growth of tree resulting in spiral grain;
- manufacturing process: sawn timber cuts with a slight angle with respect to the axis of the stem.

For visual grading, both forms of slope of grain shall be considered.

Severe slope of grain results in twisting and warping of sawn timber. Furthermore, high values of slope of grain tend to decrease the mechanical parameters such as the strength of the sawn timber. Nevertheless, weak correlation between slope of grain and strength has been found by researchers probably due to the rather seldom occurrence of severe slope of grain [21].

2.1.3. *Density and rate of growth*

Density is one of the key mechanical properties of wood and represents the third grade determining property in the strength grading process. According to EN 384 [8], the density can be measured by following two methods:

- on specimens tested to failure: the density of each specimen shall be determined on a sawn timber piece cut out close to the fracture section and free from knots and resin pockets;

- on specimens not tested to failure: the density of each specimen is determined from the ratio between the mass and volume of the test piece and divided by a coefficient equal to 1.05 in case of softwood to adjust to the density of the small defect-free pieces.

As reported by Hanhijärvi et al. [21], several researches demonstrated that density is well correlated with strength properties in case of the defect-free wood specimens. In the case of structural timber, however, the density parameter has a very large variation and only low correlations have been found in the experimental programme where the density variation in the specimens is small.

National standards specify methods of measurement of rate of growth. The Italian Standard defines the rate of growth as the average annual ring width, more specifically the ratio between a reference distance (l) and the number of annual rings (N) along the distance l. The distance l shall be identified as a straight line normal to the growth rings and either having a length of 75 mm or as the longest line normal to the growth rings. If the sawn timber contains the pith, l shall be taken outside a circle of 25 mm radius centred in the pith (**Figure 3**).

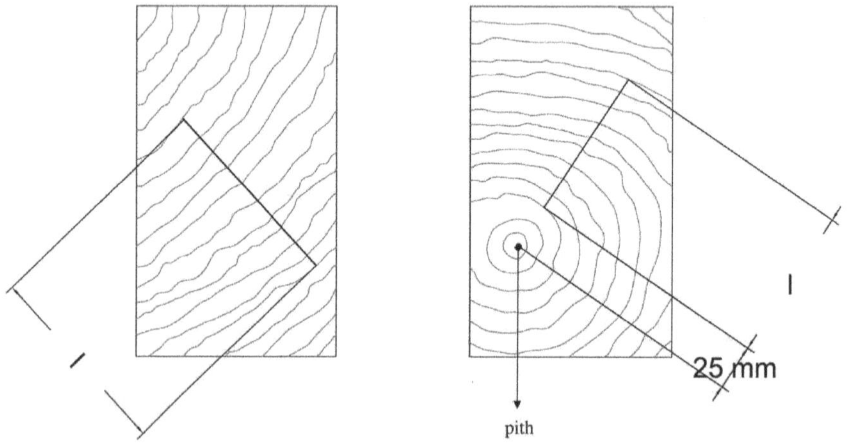

Figure 3. Annual ring width measurement according to the Italian Standard.

2.1.4. *Fissures*

Different type of fissures can occur in wood due to natural event or seasoning conditions.

Standing timber shows generally cracks or fissures, confined to the interior part of the trunk, due to a separation of fibres along the grain.

Checks, splits and shakes (**Figure 4**) generally occur during the drying process: the change in moisture content causes a variation in the volume and the occurrence of internal stresses which cause the separation of the fibres [22].

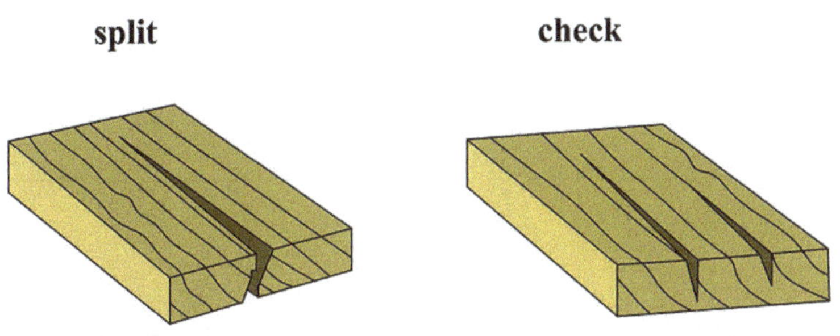

Figure 4. Fissures in sawn timber.

Checks are fissures that occur along the grain and do not extend through the sawn timber from one face to the other, while splits are fissures that extend through the sawn timber from one face to another. Shakes are cracks caused by the separation of the fibres along the annual ring growth that occurs in standing or fallen tree, or during seasoning process.

Limitation of fissures dimensions is given by grading national standards. Shakes are generally limited by grading national standards because they permit entrance of moisture, which may result in decay. For example, according to the Italian Standard, timber with shakes cannot be graded and must be rejected [23].

2.2. Geometrical characteristics

2.2.1. Wane

Wane should be restricted in squared shape planks used in buildings. Although not primarily reducing strength and stiffness, nevertheless wane can influence the practical use and further processing of the sawn timber [24]. Wane can be particularly undesirable when nail plates or connectors are used or there is transverse compression. The current harmonized standards do not cover timber with non-rectangular cross section [25] although define limits for wanes.

According to EN 1310 [14], the wane can be expressed either as a percentage of the total length of the board, measuring the length of the wane on one edge (and adding the different lengths if the plank shows more than one wane) or as a decimal fraction of the width of the edge reduced by the wane and the full width.

Wane should not be greater than one-third of the full edge and/or face [6]. German Standard DIN 4074-1 [13] defines the wane parameter, k, as the ratio between the net and the full edge of the rectangular section; the maximum permitted values can vary with the visual strength class of timber, as illustrated in **Table 1**.

According to the Italian Standard UNI 11035-1 [26], the magnitude of the wane is expressed by the ratio (s) of the projection of the wane on one side to the side length itself. Its values should be limited to 1/3 for strength class S2 and S3; a reduced value of 1/4 is allowed for class S1 [11].

Visual strength classes	Wane parameter k
S7, S7k	≤1/4
S10, S10k	≤1/4
S13, S13k	≤1/5

Table 1. Wane parameter values according to DIN 4074-1 [13].

As reported by Arriaga Martitegui et al. [27], the Spanish Standard suggests the measurements of the length of the wane and its dimensions on the edge and face of the sawn timber. The wane is evaluated as the ratio between the waneless dimension and the dimension of the rectangle into which the section fits. In length-wise direction, wane is determined as the ratio between its length and the total length of the sawn timber. Furthermore, according to Montero et al. [28], the Spanish Standard limits the maximum wane length to 1/3 of the length of the plank for *Pinus sylvestris* L. timber with thickness >70 mm. The relative dimension should also not be >1/3.

Figure 5 shows the width of wane measurements according to EN 1310 [14], German [13], Italian [11] and Spanish [12] Standards. The limitations of wane size reported in national standards on visual strength grading lead to high number of rejected sawn timber. The major effect of wanes on timber elements is related to a reduction in the cross-sectional area and as a consequence a reduction in the total load-carrying capacity. The change of shape from square to circular cross section does not affect the bending strength if the area of the sections is the same [27].

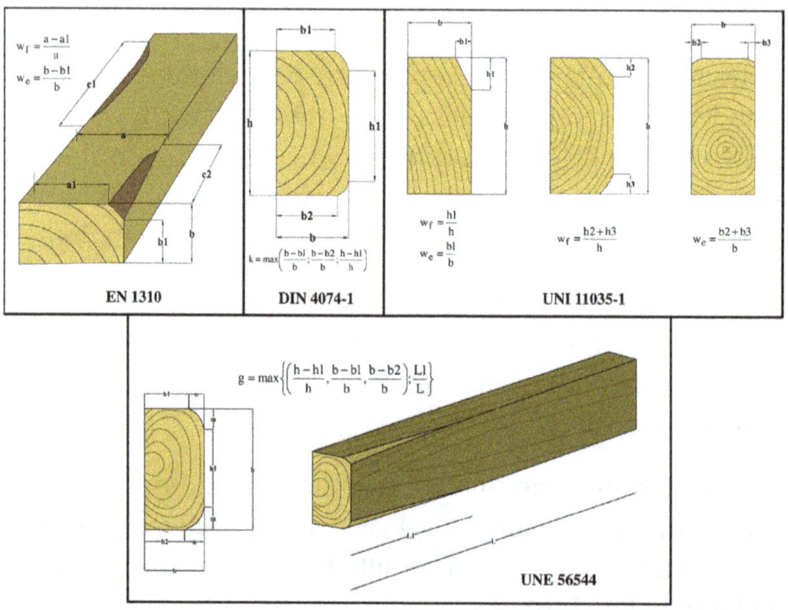

Figure 5. Width of the wane according to European visual strength grading standards (w_e = wane on edge, w_f = wane on face).

2.2.2. *Warp*

Like wanes, warps can influence the practical use and further processing of the sawn timber and thus should be restricted. The maximum distortion (spring and bow) from the straight configuration should be referred to a length of 2 m and should be measured as in the following:

- for pieces up to a length of 2 m, with reference to a straight line, expressing the result in millimetres;
- for pieces longer than 2 m, over a 2 m length, using a 2-m-long rigid straight edge applied against the piece symmetrically at the point of maximum distortion, visually estimated.

The result is expressed in millimetres per 2 m (**Figure 6**).

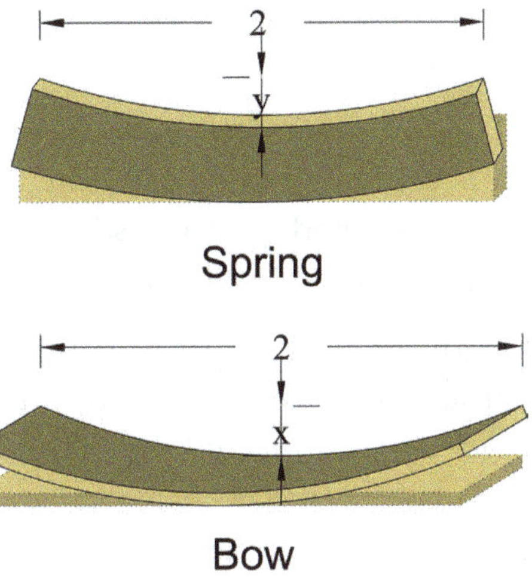

Figure 6. Spring and bow measurements.

Cup is the maximum distortion along the width of the piece, expressed as a percentage of the width (**Figure 7**).

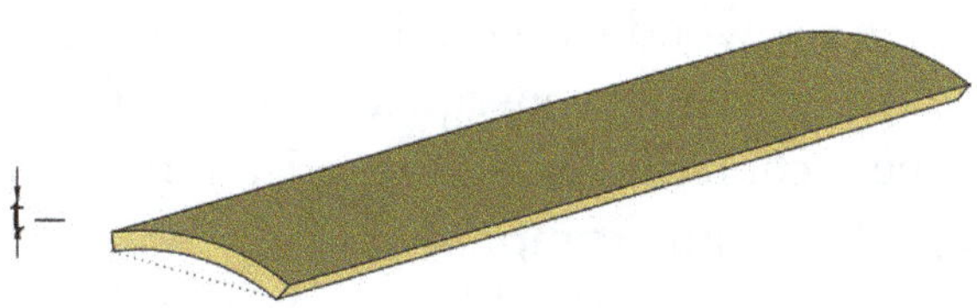

Figure 7. Cup measurements.

Twist represents the maximum distortion of the surface over a representative 2 m length and should be expressed in millimetres or as a percentage of the length of the piece (**Figure 8**).

According to EN 14081-1 [6], for both visual and machine graded structural timber, maximum warp over 2 m of length of the board should be limited to the values listed in **Table 2**.

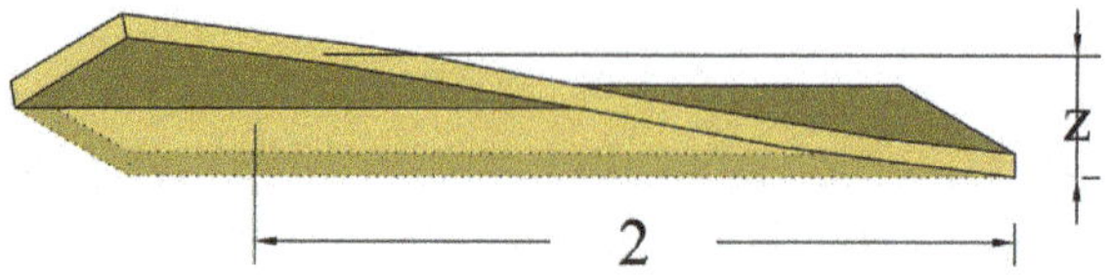

Figure 8. Twist measurement.

Warp	Strength classes	
	Lower than or equal to C18, D18, T11	**Greater than C18, D18, T11**
Bow	20 mm	10 mm
Spring	12 mm	8 mm
Twist	2/25 mm width	2/25 mm width
Cup	Unrestricted	Unrestricted

Table 2. Warp limits according to EN 14081-1 [6].

Warp depends upon the moisture content of timber and can therefore change with time, and thus, limits in **Table 2** should be considered only in case of dry grading. Longitudinal curvature in square section pieces may be assessed using the limits for bow [6].

Sandberg [29] studied the influence of repeated cycles of wetting and drying in terms of warps on sawn timber of Pine and Spruce and stated that warp and the number of cracks increases if timber undergoes repeated cycles of wetting and drying.

2.3. Biological characteristics

Biological organisms such as fungi, bacteria and insects may attack and damage wood. Four critical elements must be present for the wood to be damaged by biological organisms: temperature, moisture, oxygen and a food source.

Fungi require all the four critical elements to be present for attacking; however, the most important one is the presence of moisture in the form of free water. In general, an infection of fungi leads to a reduction in the wood structural integrity. Cross-sectional and mechanical strength reductions are the two principal consequences of the fungi infestation in wood elements: a 10% reduction in the section dimensions may lead up to 50% reduction in mechanical properties of wood. Impact strength, compression perpendicular and parallel to grain are the most affected mechanical properties. Moulds and staining fungi generally affect only the impact strength of wood and do not cause a reduction in the section dimensions. Soft and white rot fungi are most common in hardwoods such as Aspen, while brown rot fungi generally attack softwoods such as Pines, Firs and Spruces. These types of fungi cause degradation and affect the mechanical strength of wood [30].

In general, the biotic decay caused by fungi or insect attacks leads to a reduction in the cross section of the wood elements. For this reason in case of grading new timber, the members subjected to a biotic decay must be rejected. Nevertheless, when grading timber members belonging to ancient and historical buildings, the presence of decayed elements is inevitable [23].

2.4. Other characteristics

Reaction wood is the term generally used for describing the abnormal tissue of wood, which is called compression wood in softwood and tension wood in hardwood. In general reaction, wood is characterized by higher density if compared to normal wood: 7% greater in tension wood and 35% greater in compression wood. Several defects, such as warps and surface checks, are caused by the presence of compression wood in timber elements during the drying process. Moreover, a brittle failure appears in timber containing compression wood [31, 32].

The influence on the mechanical properties of both compression and tension wood compared to normal wood has been extensively studied: as reported by Wimmer and Johansson [33], compression wood is characterized by higher values of density and lower modulus of Young, bending and tensile strengths. Higher values of density, modulus of Young, bending strength and tensile strength are achieved in tension wood compared to normal wood.

National visual strength grading rules limit the amount of compression wood in softwood elements, while no limits are indicated for tension wood in hardwood elements. The amount of reaction wood according to UNI 11035-1 [26] is given by the ratio between the sum of the widths of strips containing the reaction wood and the perimeter of the cross section as shown in **Figure 9**.

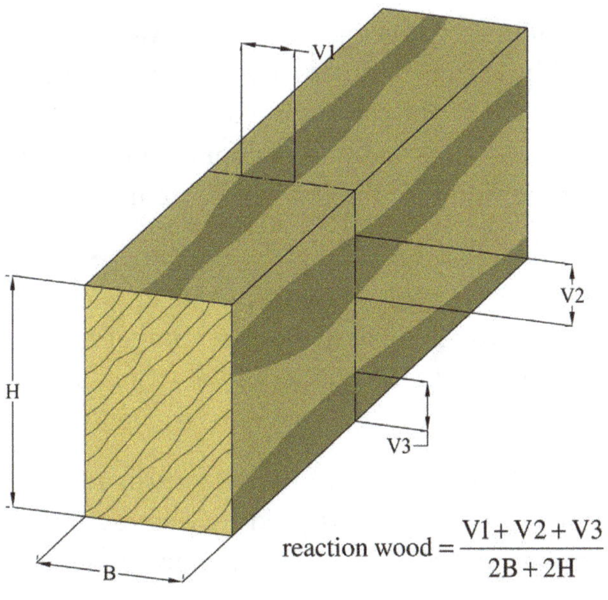

Figure 9. Reaction wood measurement.

3. Machine grading

Machine grading has been commonly used in a number of countries for over 40 years. Like visual grading, the machine grading sorts the sawn timber into strength classes by means of some non-destructive measurements related to the mechanical properties. According to the European Standard EN 14081-1 [6], rectangular cross-sectional timber should be sorted into strength categories based on strength, stiffness and density, as well as some geometrical characteristics that should be limited because of their potential strength-reducing effects.

The difference between machine grading and visual grading is that a machine can predicts the grade of timber by measuring some non-destructive properties, usually known as IPs (indicating properties). The IPs are measurements or combination of measurements taken by the grading machine that are closely related to one or more of the grade determining properties [34].

Acceptance criteria are formulated in terms of intervals for the corresponding IP that have to be matched to qualify a piece of timber to a certain grade [35].

These IPs are usually more powerful predictors of quality with respect to those measured by visual grading, and the grading can be done at a much faster rate with less risk of human error [25]. The oldest grading machines measured the modulus of elasticity of timber via non-destructive bending tests. Nowadays, new technologies and a great variety of IPs are used, with good correlations with the wood properties: for example, ultrasonic pulse velocity and longitudinal or flexural resonant frequency are measured in order to determine the dynamic modulus of elasticity, and X-ray analyses are performed to determine density and identify knots, etc.

Two basic machine grading systems are provided by the European Standard EN 14081-2 [34]: the "output-controlled" and the "machine-controlled" methods.

The "output-controlled" system was developed in North America. The control is based on statistical procedures assessed by means of destructive testing on specimens randomly selected from the daily production. Based on the correlation between the IP and the grade determining property (e.g. strength, stiffness), this method examines the output of the grading machine continuously by observing the values of the IPs which are measured by the grading machine non-destructively [36].

In this way, machine settings are monitored and can be adjusted after each test in order to optimize the prediction of the properties of the graded timber material [37]. This implies that machine settings are strictly related to the quality of the wood, and the same type of machine could have non-identical performance.

On the other hand, the output control system has been proved to be very expensive since a large amount of sawn material has to be assessed by destructive tests, and not all the data from these tests can be used for the recalibration of the grading machine [35].

Thus, the "output-controlled" system is suitable in sawmills grading having production limited in sizes, species and grades because of the need of a continuous check of the grading process. The output control procedure currently requires only a verification of the bending strength and the bending stiffness. The measure of density is not required from the standardized control procedure [38, 39].

The "machine-controlled" system was developed in Europe. Due to the large number of sizes, species and grades, it was not possible to carry out quality control tests on timber specimens taken from production [34]. Machine settings derived from the results of destructive testing programmes have been developed in order to have the same settings for the same machine types. Machine-controlled systems are not based on specific measurements, but on the capability of the machine to assign any piece of timber to a specific grade on the condition that the required characteristic values of the assigned grade have been satisfied [39]. For this reason, modern grading machines are based on non-destructive testing, contact-free measures or on their combination. Several authors demonstrated good correlations among non-destructive parameters and mechanical and stiffness properties of timber (ultrasounds measures or vibration methods) [39] or density (X-ray measures) [40–42]. Some models of grading machines incorporated a contact-free in-line moisture metre, so stiffness and density measures are automatically adjusted to the reference conditions (12%) [39].

The effectiveness of a grading machine depends on the speed and on its capacity to subdivide the ungraded timber into sub-classes of graded timber in order to satisfy some predefined requirements [35]. The relationship between the IP and the three grade determining properties (density, stiffness and strength) varies with the wood species and with the region of provenience. Contrary to the output control system, the machine-controlled system is based on settings that are unique for grading region and wood species [25]. Thus, grading machines of the same type have the same settings if installed in the same region for grading the same wood species. However, both machine and output-controlled systems have revealed some problems and mainly related to the machine control strategy which is considered incapable to take into account the large scatter in the origin, sawing pattern and growth condition and other properties of the ungraded base material [35]. For this reason, both systems often require a further visual inspection in case of some strength-reducing characteristics were not automatically detected by the machine, for example, in the case of bending type machines, where the end of the pieces cannot be graded completely and a further visual inspection is necessary. EN 14081-1 [6] requires also some visual characteristics to be checked for each piece (warps, wane, fissures, insect damage, etc.).

4. Visual strength grading of Sardinian Maritime Pine

In this section, the results of an experimental programme aimed at identifying the visual strength grades of Sardinian timber are discussed and analysed.

According to the National Forest Inventory (INSC) [43], one-fourth of Sardinia is covered by wood and about 5000 hectares are covered by conifers, in particular Stone Pine (*Pinus pinea* L.), Aleppo Pine (*P. halepensis* Mill.), Corsican Pine (*P. nigra* Arn.), Maritime Pine (*P. pinaster* Ait.) and Radiata Pine (*P. pinaster* D. Don) [1].

Among the conifers, visual strength grading methodology applied to Sardinian Maritime Pine is reported and discussed. This species is quite widespread also in other Mediterranean regions such as the Iberian Peninsula, France, Corsica, etc. and is relatively fast growing.

Three different growth regions, one located in the northern part, one in the centre and another in the southern part of Sardinia, were chosen in order to satisfy two requirements:

- density of population higher than 800 plants/ha;
- stem bark size higher than 18 cm.

The experimental programme was carried out on about 300 boards, 3.00 m long, 0.035 m thick and 0.125 m wide.

Visual strength grading procedure according to UNI 11035-1 [26] was applied on boards after the drying process on a climate chamber at relative humidity of 65% and 20°C of temperature until constant weight was achieved.

Table 3 gives a statistical summary of the most problematic geometrical and morphological parameters for visual strength grading and their values into the S1, S2, S3 and rejected (R) visual grades [11].

As shown in **Table 3**, about 50% of boards were rejected and could not be included into visual grades due to three parameters: knot, knot cluster and twist.

Parameter		S1	S2	S3	Rejected
Sample	Number [%]	5	18	30	47
Knot parameter	AV	0.18	0.25	0.32	0.4
	St.Dev	0.06	0.07	0.16	0.14
	CoV [%]	33.33	29.28	50.24	36.22
Knot cluster	AV	0.3	0.42	0.62	0.76
	St.Dev	0.1	0.09	0.11	0.12
	CoV [%]	33.33	21.55	18.62	16.10
Twist	AV [mm]	11.07	14.19	14.7	21.37
	St.Dev [mm]	3.24	3.08	8.74	9.03
	CoV [%]	29.29	21.7	59.53	42.25
Bow	AV [mm]	5.50	6.33	7.4	8.92
	St.Dev [mm]	2.9	3.38	4.01	6.16
	CoV [%]	57.75	53.37	54.30	69.02
Spring	AV [mm]	4.0	5.04	6.05	5.61
	St.Dev [mm]	1.15	2.5	4.66	4.06
	CoV [%]	28.87	49.63	77.01	72.42

AV, average; St.Dev, standard deviation; CoV, coefficient of variation.

Table 3. Geometrical and morphological parameters.

Sardinian Maritime Pine timber is a low-quality wood: only 5% of the overall sample belongs to S1 visual grade, while about 45% of boards are divided into S2 and S3 visual grades.

Maritime Pine boards are characterized by high scatter of all the geometrical and morphological parameters and by high values of knot, knot cluster and twist warping. **Figure 10** shows the distribution of visual grades according to each of these parameters.

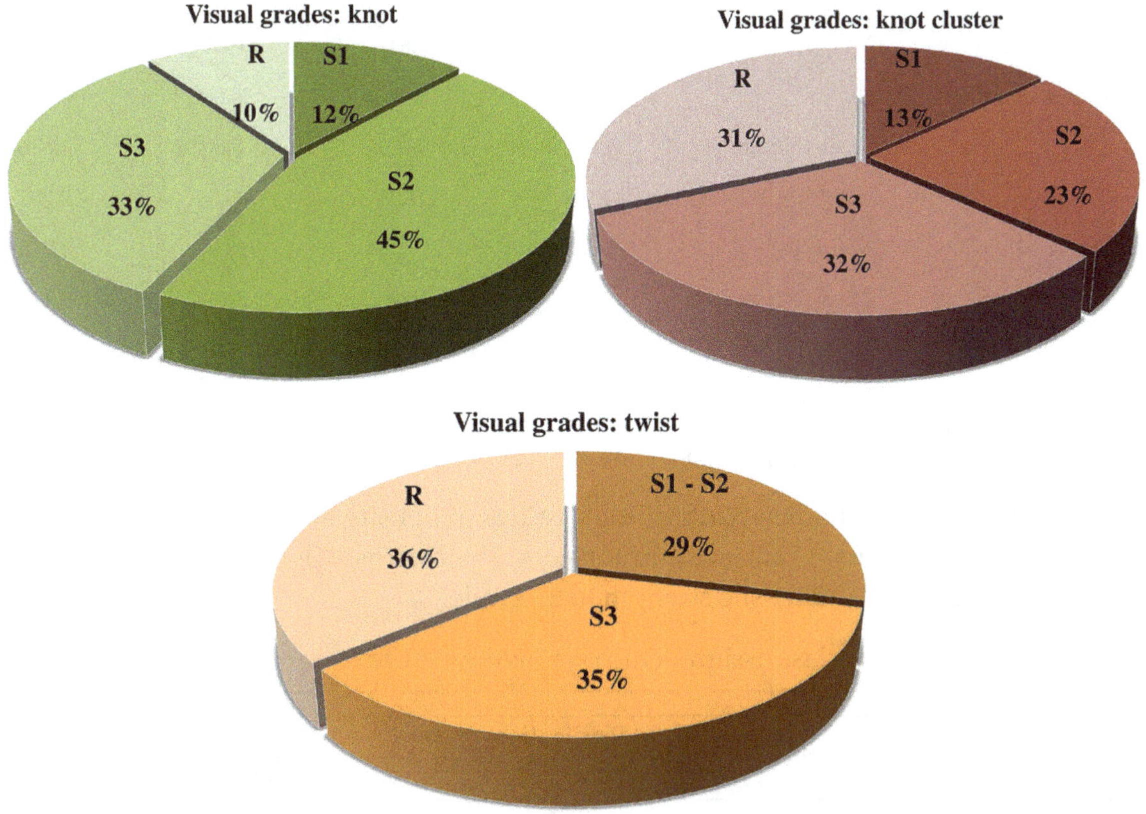

Figure 10. Distribution of visual grades according to knot, knot cluster and twist parameters.

Only about 10% of boards were rejected due to knot parameter, while more than 30% were rejected considering both knot cluster parameter and twist warping.

The analysis of the visual grades distributions of knot and knot cluster parameters and twist warping highlights the low quality of wood: the boards that are included into the highest visual grade S1 are about 10% due to knot parameters, while more than 70% of boards are included into R and S3 grades.

According to UNI 11035-2 [11], the boards were subdivided into visual grades, and then, they were tested to destruction in order to evaluate the characteristic values of density, modulus of elasticity and tensile strength and to determine which strength class is satisfied by the visual grades.

Tensile tests were carried out in the worst sections of the boards for measuring the static elastic modulus of elasticity and the failure load according to EN 408 [9].

Table 4 shows the statistical summary of the three keys parameters (density, modulus of elasticity in tension and tensile strength) used for determining the strength classes according to EN 338 [7].

Parameter		S1	S2	S3	Rejected
Density	[kg/m³]	520.10	506.69	501.19	504.95
	St.Dev [kg/m³]	69.24	38.71	53.96	47.68
	CoV [%]	13.31	7.64	10.77	9.44
Modulus of elasticity in tension	[N/mm²]	9208.50	8196.50	8584.16	4387.32
	St.Dev [N/mm²]	1233.48	962.06	1480.95	1246.70
	CoV [%]	20.09	17.39	25.56	22.02
Tensile strength	[N/mm²]	14.19	13.51	12.96	11.85
	St.Dev [N/mm²]	3.08	4.20	3.26	3.17
	CoV [%]	21.70	31.06	25.16	26.73

St.Dev = standard deviation; CoV = coefficient of variation.

Table 4. Sardinian Maritime Pine: density, modulus of elasticity in tension and tensile strength.

All the visual grades are characterized by high values of density and low values of both tensile strength and modulus of elasticity in tension. The boards belonging to R grade show similar values of density and tensile strength to S3 grade.

The matrix of the strength class assignments is shown in **Table 5**.

Visual grades	Parameter	Tensile strength classes							A
		T10	T11	T12	T13	T14	T28	T30	
S1	Density							X	T11
	Modulus of elasticity in tension		X						
	Tensile strength					X			
S2	Density						X		T10
	Modulus of elasticity in tension	X							
	Tensile strength				X				
S3	Density						X		T10
	Modulus of elasticity in tension	X							
	Tensile strength			X					
R	Density						X		<T10
	Modulus of elasticity in tension								
	Tensile strength		X						

A = assigned tensile strength class.

Table 5. Sardinian Maritime Pine: matrix of the strength class assignments.

Several considerations can be made from the matrix of strength class assignments shown in **Table 5**:

- according to density, the high values of all the visual grades are confirmed by the T-strength class assignments: S1 grade corresponds to T30, while S2, S3 and R grades belong to T28 strength class;
- the low values of modulus of elasticity cause a downgrade in the strength class assignments for all the visual grades;
- the maximum T-class assignment according to tensile strength corresponds to T14 strength class, while T11 is achieved by the modulus of elasticity in tension.

According to the three key parameters, both S2 and S3 Sardinian Maritime Pine timber visual grades can be assigned to T10 strength class, while S1 visual grade can be assigned to T11 strength class.

Furthermore, the R-rejected class could be assigned to T10 class for tensile strength and density like the S3 and S2 visual grades. This suggests that the visual grading rule proposed in UNI 11035-1 [26] is too conservative regarding the limits of the defectiveness parameters, and a new proposal for visually grading of Sardinian Maritime Pine timber should be developed.

5. Conclusions

Maritime Pine is a very resinous, low strength conifer. Knots and warping are amongst the worst defects and are considered a disadvantage for structural uses because they markedly affect the strength and stiffness of timber.

In addition, wood production is affected by several factors depending both on the growth area (altitude, wind, type of soil, rainfall, etc.) and on genetic factors which could result in a high variability of mechanical properties of the sawn timber [44]. For these reasons, several European countries have developed their own grading rules for locally grown species of timber based on the same criteria of the reference European Standard.

Carballo et al. [45] visually graded according to the Spanish Standard [12], destructively tested Maritime Pine sawn timber from Galicia and stated that despite the high percentage of rejected board (37%), the base material exhibited a great structural capacity corresponding to C24 and C30 strength classes according to EN 338 [7].

Morgado et al. [44] visually graded and destructively tested Maritime Pine poles from Portugal and compared test results with those obtained in similar researches. They stated that strength values obtained for Maritime Pine were significantly higher than those obtained from other species. Moreover, a proposal for visually grading Portuguese Maritime Pine roundwood was developed [46].

Several researches demonstrated the suitability of low-quality wood for the production of structural elements [47, 48] like cross-laminated timber (CLT) panels because the lamination and system effect in CLT production reduce the influence of the defects (knots, warp, etc.) of the base material. Furthermore, as reported by Concu et al. [42], preliminary results on CLT panels made of Sardinian Maritime Pine wood confirmed that medium quality panels can be produced and used as horizontal and vertical elements in civil engineering structures.

In conclusion, low-quality wood as Maritime Pine must be graded and classified into strength classes based on strength, stiffness and density before any use as a structural element. Grading rules for locally grown species should be drawn in order to minimize the negative effect of warping and other geometrical characteristics on strength class assignments.

Author details

Trulli Nicoletta[1*], Monica Valdés[1], Barbara De Nicolo[1] and Massimo Fragiacomo[2]

*Address all correspondence to: ntrulli@unica.it

1 Department of Civil Engineering, Environmental and Architecture, University of Cagliari, Cagliari, Italy

2 Department of Civil, Construction-Architecture and Environmental Engineering, University of L'Aquila, Italy

References

[1] Fragiacomo M., Riu R., Scotti R. Can structural timber foster short procurement chains within Mediterranean forests? A research case in Sardinia. South-east European Forest. 2015;**6**(1):107–117. doi:10.15177/seefor.15-09

[2] Buchanan A. Energy and CO_2 advantages of wood for sustainable buildings. NZ Timber Design Journal. 2007:**15**(1):11–21.

[3] Buchanan A., Honey B. Energy and carbon dioxide implications of building constructions. Energy and Building. 1994;**20**:205–217.

[4] Oliva G.A., Merino B.V., Seco F.G.J.I., Conde García M., Hermoso Prieto E. Effect of growth conditions on wood density of Spanish *Pinus nigra*. Wood Science and Technology. 2006;**40**:190–204. doi:10.1007/s00226-005-0014-0

[5] REGULATION (EU) (2011) No 305/2011 of 9 March 2011 laying down harmonized conditions for the marketing of construction products and repealing Council Directive 89/106/EEC.

[6] EN 14081-1 (2016). Timber Structures. Strength graded structural timber with rectangular cross section. Part I: General requirements.

[7] EN 338 (2016). Structural timber—Strength classes.

[8] EN 384 (2016). Structural timber—Determination of characteristic values of mechanical properties and density.

[9] EN 408 (2010) Timber structures—Structural timber and glued laminated timber—Determination of some physical and mechanical properties.

[10] EN 1912 (2007). Structural timber—Strength classes—Assignment of visual grades and species.

[11] UNI 11035-2 (2010). Visual strength grading for structural timbers. Part 2: Visual strength grading rules and characteristics values for structural timber population.

[12] UNE 56544 (2011). Visual grading for structural sawn timber. Coniferous timber.

[13] DIN 4074 (Part 1–5) (2003). Strength grading of wood. Part 1: coniferous sawn timber.

[14] EN 1310 (1997). Round and sawn timber. Method of measurements of features.

[15] Machado J.S., Sardinha R.A., Cruz H.P. Feasibility of automatic detection of knots in maritime pine timber. Wood Science and Technology. 2004;**38**:277–284. doi:10.1007/s00226-004-0224-x

[16] Koman S., Feher S., Abraham J., Tascner R. Effect of knots on the bending strength and the modulus of elasticity of wood. Wood Research. 2013;**58**(4):617–626.

[17] Olsson A., Oscarsson J., Serrano E., Källsner B., Johansson M., Enquist B. Prediction of timber bending strength and in-member cross-sectional stiffness variation on the basis of local wood fibre orientation. European Journal of Wood and Wood Products. 2013;**71**:319–333. doi:10.1007/s00107-013-0684-5

[18] Krzosek S. Timber strength grading of *Pinus sylvestris* L. using a visual method according to Polish standard PN-82/D-94021 and German standard DIN 4074. Wood Research. 2011;**56**(3):435–440.

[19] Adell Almazán F.J., Hermoso Prieto E., Arriaga Martitegui F., Richter C. Comparison of the Spanish visual strength grading standard for structural sawn timber (UNE 56544) with the German one (DIN 4074) for Scots pine (*Pinus sylvestris* L.) from German. Holz Roh Werks. 2008;**6**:253–258. doi:10.1007/s00107-008-0241-9

[20] Stapel P., van de Kuilen J.W.G., Strehl O. Visual strength grading in Europe. In: International Council for Research and Innovation in Building and Construction; CIB, Working Commission W18 -Timber Structures, Meeting 45, August 2012; Växjö, Sweden. CIB-W18/45-5-2.

[21] Hanhijärvi A., Ranta-Maunus A., Turk G. Potential of strength grading of timber with combined measurement techniques, Report of the Combigrade project—phase 1, VTT Publications 568. 2005.

[22] Esteban M., Arriaga F., Íñiguez G., Bobadilla I., Mateo R. The effect of fissures on the strength of structural timber. Materiales de Construcción. 2010;**60**(299):115–132. doi:10.3989/mc.2010.48208

[23] Piazza M., Riggio M. Visual strength-grading and NDT of timber in traditional structures. Journal of Building Appraisal. 2008;**3**(4):267–296. doi:10.1057/jba.2008.4

[24] Schickhofer G., Augustin M. Final report Project INTELLIWOOD—Working package 3—Strength correspondence. 2001.

[25] Ridley-Ellis D., Stapel P., Bano V. Strength grading of sawn timber in Europe: an explanation for engineers and researchers. European Journal of Wood and Wood Products. 2016;**74**:291–306. doi:10.1007/s00107-016-1034-1

[26] UNI 11035-1. (2010). Visual strength grading for structural timbers. Part I: Terminology and measurements of features.

[27] Arriaga Martitegui F., Esteban Herrero M., Argüelles Álvarez R., Bobadilla Maldonado I., Íñiguez González G. The effect of wane on the bending strength of solid timber beams. Materiales de Construcción. 2007;**57**(288):61–76.

[28] Montero M.J., Mateo R., Íñiguez-González G., Arriaga-Martitegui F., Hermoso E., Esteban M. Visual grading of large cross section structural timber of *Pinus sylvestris* L. according to UNE 56544:2007 standard. In: SHATIS'11 International Conference on Structural Health Assessment of Timber Structures; June 2011; Lisbon Portugal. 2011.

[29] Sandberg D. Radially sawn timber. The influence of annual ring orientation on crack formation and deformation in water soaked pine (*Pinus sylvestris* L) and spruce (*Pinea abies* karst) timber. Holzals Roh-und Werkstaff. 1997;**55**(3):175–182.

[30] Clausen C.A. Biodeterioration of wood. In: Department of Agriculture, Forest Service, Forest Products Laboratory, editors. Wood Handbook—Wood as an Engineering Material. General Technical Report FPL-GTR-190. Madison, WI: U.S:2010. pp. 14/1–14/16.

[31] Porteous J., Kermani A., editors. Structural Timber Design to Eurocode 5. 2nd ed. Wiley-Blackwell; 2013. 640 p. Oxford. United Kingdom.

[32] Tarmian A., Remond R., Faezipour M., Karimi A., Perré P. Reaction wood drying kinetics: tension wood in *Fagus sylvatica* and compression wood in *Picea abies*. Wood Science and Technology. 2009;**43**:113–130. doi:10.1007/s00226-008-0230-5

[33] Wimmer R., Johansson M. Effects of Reaction Wood on the Performance of Wood and Wood-Based Products. In: Gardiner B., Barnett J., Saranpää P., Gril J., editors. The Biology of Reaction Wood. Springer Series in Wood Science, Springer-Verlag Berlin Heidelberg; 2014. pp. 225–248. doi:10.1007/978-3-642-10814-3_8

[34] EN 14081-2:2010+A1:2012. Timber structures—Strength graded structural timber with rectangular cross section. Part 2: Machine grading; additional requirements for initial type testing.

[35] Deublein M., Steiger R., Köhler J. Quality Control and Improvement of Structural Timber. In: Final Conference of COST E53 "The Future of Quality Control for Wood & Wood Products"; 4-7th May 2010; Edinburgh. 2010.

[36] Sandomeer M.K., Köhler J., Faber M.H. Probabilistic Output Control for Structural Timber—Fundamental Model Approach. In: CIB-W18: International Council for Research and Innovation in Building and Construction—Working Commission W18—Timber Structures. Meeting forty-one; August 2008; St. Andrews, Canada. 2008. CIB-W18/41-5-1.

[37] Deublein M., Köhler J., Linsenmann P. The efficient control of grading machine settings. In: CIB-W18: International Council for Research and Innovation in Building and Construction—Working Commission W18—Timber Structures. Meeting forty; August 2007; Bled, Slovenia. 2007. CIB-W18/40-5-2.

[38] Kovryga A., Stapel P., van de Kuilen J.V.G. Safety of timber—an analysis of quality control options. In: WCTE 2014: World Conference on Timber Engineering; 10–14th August; Quebec City, Canada. 2014.

[39] Bacher M. Comparison of different machine strength grading principles. In: Conference COST E53; 29–30th October; Delft, The Netherlands. 2008.

[40] Nocetti M., Bacher M., Brunetti M., Crivellaro A., van der Kuilen J.W.G. Machine grading of Italian structural timber: preliminary results of different wood species. In: WCTE 2010: World Conference on Timber Engineering; 20-24th June; Trentino, Italy. 2010.

[41] Concu G., De Nicolo B., Trulli N., Valdés M., Fragiacomo M. Strength class prediction of Sardinia grown timber by means of non destructive parameters. Advanced Materials Research. Trans Tech Publications. Switzerland. 2013; **778**:191–198. doi:10.4028/www.scientific.net/AMR.778.191

[42] Concu G., De Nicolo B., Valdés M., Fragiacomo M., Menis A., Trulli N. Experimental grading of locally grown timber to be used as structural material. In: Chang, Al Bahar, Zhao, editors. Advances in Civil Building Materials. CEBM 2012: 2nd International Conference on Civil Engineering and Building Materials; 17–18th November 2012; Hong Kong. London: Taylor and Francis Group; 2013. pp. 189–193. ISBN: 978-0-415-64342-9.

[43] Gasparin P., Tabacchi G., editors. INFC 2005: L'Inventario Nazionale delle Foreste e dei serbatoi forestali di Carbonio. Secondo inventario forestale nazionale italiano. Metodi e risultati. Ministero delle Politiche Agricole, Alimentari e Forestali; Corpo Forestale dello Stato. Consiglio per la Ricerca e la Sperimentazione in Agricoltura, Unità di ricerca per il Monitoraggio e la Pianificazione Forestale. Bologna: Il Sole 24 ore Edagricole; 2011. 653 p. ISBN-10: 8850653948. (in Italian).

[44] Morgado T.F.M., Rodrigues J.N.A., Saporiti J., Dias A.M.P.G. Grading and testing of Maritime pine and larch roundwood. In: Conference COST E53; 29-30th October; Delft, The Netherlands.2008.

[45] Carballo J., Hermoso E., Fernández-Golfín J.I. Mechanical properties of structural maritime pine sawn timber from Galicia [Spain] (*Pinus pinaster* Ait. ssp. atlantica). Investigación agraria. Sistemas y recursos forestales. 2009;**18**(2):152–158. doi:10.5424/fs/2009182-01058

[46] Morgado T.F.M., Saporiti Machado J., Dias A.M.P.G., Cruz H., Rodrigues J.N.A. Grading and Testing of Maritime Pine Roundwood. In: WCTE 2010: World Conference on Timber Engineering; 20–24th June; Trentino, Italy. 2010.

[47] Smith R.E. Interlocking Cross-Laminated Timber: alternative use of waste wood in design and construction. In: Building Technology Educators' Society, editor. BTES Conference 2011—Convergence and Confluence; 4–7th August; Toronto, Ontario, Canada.

[48] Negri M., Gravić I., Marra M., Fellin M., Ceccotti A. Using low quality timber for X-Lam: raw material characterisation and structural performance of walls under semi-dynamic testing. In: WCTE 2012: World Conference on Timber Engineering; 16–19th July; Auckland, New Zealand. 2012.

8

Experimental Analyses and Numerical Models of CLT Shear Walls under Cyclic Loading

Valeria Awad, Linda Giresini, Mikio Koshihara,
Mario Lucio Puppio and Mauro Sassu

Abstract

This paper reports the results of an experimental campaign performed at the University of Tokyo on cross-laminated timber (CLT) panels subjected to lateral loads. Analytical and numerical interpretations are provided as well, comparing the experimental analysis results with two methods: firstly, an analytical method to preliminarily evaluate the ultimate strength of the four panels, based on the geometrical dimensions of the openings and of the panel; secondly, a finite element model has been developed in order to provide some guidelines for calculating the stiffness and elastic behaviour of CLT panels subjected to lateral loads. The experimental tests showed that the CLT panels are as more brittle and stiffer as more the difference between the total panel area and the fenestrated area is high. The presence of large openings determined stress concentration at the corners where failure occurred for the attainment of the maximum tension strength in the inner layer. The proposed analytical formulation was shown to fairly closely predict the ultimate strength of panels with same geometry, characteristics and boundary condition, allowing preliminary information of this relevant parameter.

Keywords: cross-laminated timber, CLT, cyclic tests, shear walls, cut-out openings, FE model

1. Introduction

Cross-laminate timber, also identified as CLT or X-lam, is a relatively new technology widely used in Europe since the early 2000s. In the last few years, the use of CLT system in countries such as United States, Canada and New Zealand has improved bringing hundreds of impressive buildings and showing that timber constructions, together with other natural materials [1], can be competitive, particularly for mid-rise and high-rise buildings. However, no specific design codes for CLT constructions were provided in Europe while Canada and United States have only recently published a CLT Handbook (US and Canadian version [2]) that provides key technical information related to manufacturing, design and performances of CLT, providing support for design and construction of CLT systems as alternative solutions in building codes. In Europe, Eurocode 5 governs the design of timber structures [3] and Eurocode 8 gives indications specifically related to seismic design [4]. In the literature, many contributions offer different interpretations of CLT panels subjected to lateral loads [5–8]. Nevertheless, in Japan, where timber buildings cover 50% of the constructions, a practical structural design of CLT buildings is not yet issued. To establish the law on the design method of CLT constructions suited with the Japanese regulations of seismic design, some research projects on CLT structures have been recently performed.

In the research field of CLT panels, an important contribute was given by the University of Ljubljana, Slovenia, where numerous quasi-static monotonic and cyclic tests were carried out on walls with lengths of 2.44 and 3.2 m and a height of 2.44 or 2.72 m [9]. Wall panels were subjected to gravity load induced by ballast as constant vertical load and a displacement-controlled hydraulic actuator as driver of the cyclic horizontal load. Moreover, for all the panels, three different cases of boundary conditions, from the cantilever type to the pure shear, were applied. These experiments have confirmed the importance of boundary conditions and the type of horizontal loading [10]. At a later stage, Dujic et al. [11–13] carried out other experimental and parametric studies to estimate the racking strength and stiffness of CLT wall panels with openings. Models were numerically tested by running non-linear static pushover analysis and results of calculations were compared with the test results. The experience conducted by Dujic et al. showed that cross-laminated panel with many openings has lower shear stiffness, but load-bearing capacity is not reduced as much, because failures are mostly concentrated in anchoring areas and in corners around openings with smashing and tearing of wood. Results were used to develop an exact mathematical model describing the relationship between the shear strength and stiffness of CLT wall panels without opening and panels with certain area of openings. An experimental study to quantify the seismic behaviour of X-Lam wall panels subjected to lateral loads was performed at the CNR-IVALSA Italian Institute [14]. The single-wall panels (2.95 × 2.95 m) included walls with openings and without openings and different connections layout and subjected to different levels of vertical loads. The results of these tests showed that the connections have a dominant role in the overall behaviour of the wall. As described in the Canadian edition of the CLT Handbook [2], in 2010, the FPInnovations Research Institute, in Vancouver, carried out 32 monotonic and cyclic tests [15]. The tests were conducted for 12 different configurations employing different wall-to-wall and wall-to-floor connections. As connectors, hold-down and brackets connectors were used, while as fastener, ring nails, spiral nails, self-threaded screws and timber rivets were utilized. Results of the experiments showed that CLT walls can have adequate seismic performance especially when hold-downs with nails on each end are used.

This paper deals with the interpretation of experimental tests carried out on full-size CLT panels with cut-out openings. The cyclic tests here described are performed on four three-layered 90-mm thick shear walls (30-30-30 mm) characterized by similar geometry, same boundary conditions and different cut-out openings. The experimental analyses results, performed up to timber and connection failures, were then compared with two methods. Firstly, an analytical method to preliminarily evaluate the ultimate strength of the four panels, based on the geometrical dimensions of the openings and of the panel, is proposed. Additionally, a finite element model has been developed in order to provide some guidelines for calculating the stiffness and elastic behaviour of CLT panels subjected to lateral loads.

2. Experimental tests on CLT three-layered panels

The experimental tests, carried out in the laboratories of the University of Tokyo, were addressed to specific aims. The first one was related to evaluating general performances of lateral-loading tests on four types of cross-laminated shear wall systems with openings. Additionally, effort was made to investigate the panels' response under cyclic loading, considering the influence of wood and connection features.

2.1. CLT panels and test set-up

The four test specimens of the full-scale CLT shear walls have openings (doors or windows, **Table 1**) and are three-layered 90-mm thick panels (30-30-30). The width of the one opening type is 4000 mm while the two opening panels are 6000 mm in width; their height is 2700 mm. These dimensions are displayed in **Figure 1**. From now on, the panels labelled in **Table 1** as 1S4-A, 1S4-B, 1S6-C and 1S6-D will be referred to as panels A, B, C and D, respectively. The panels are built in Sugi timber (Japanese cedar), which is characterized by a lower value of Young's modulus (4–6 GPa).

As example, panel C is displayed in **Figure 2a**. At the bottom of the wall, as shown in **Figure 2b**, two different types of connection between the panel and the steel foundation were used: the tensile connector and the shear connector. U-shaped steel elements with 12 screws each were adopted as tensile connections (**Figure 3a**) while U-shaped steel elements with 16 screws acted as shear connections (**Figure 3b**). The screws had a length of 65 mm and a diameter of 7 mm for all the joints.

Specimen	Opening type	Openings number	Length
1S4-A	Door	1	4 m
1S4-B	Window	1	4 m
1S6-C	Door	2	6 m
1S6-D	Window	2	6 m

Table 1. Labels and openings of the tested CLT panels.

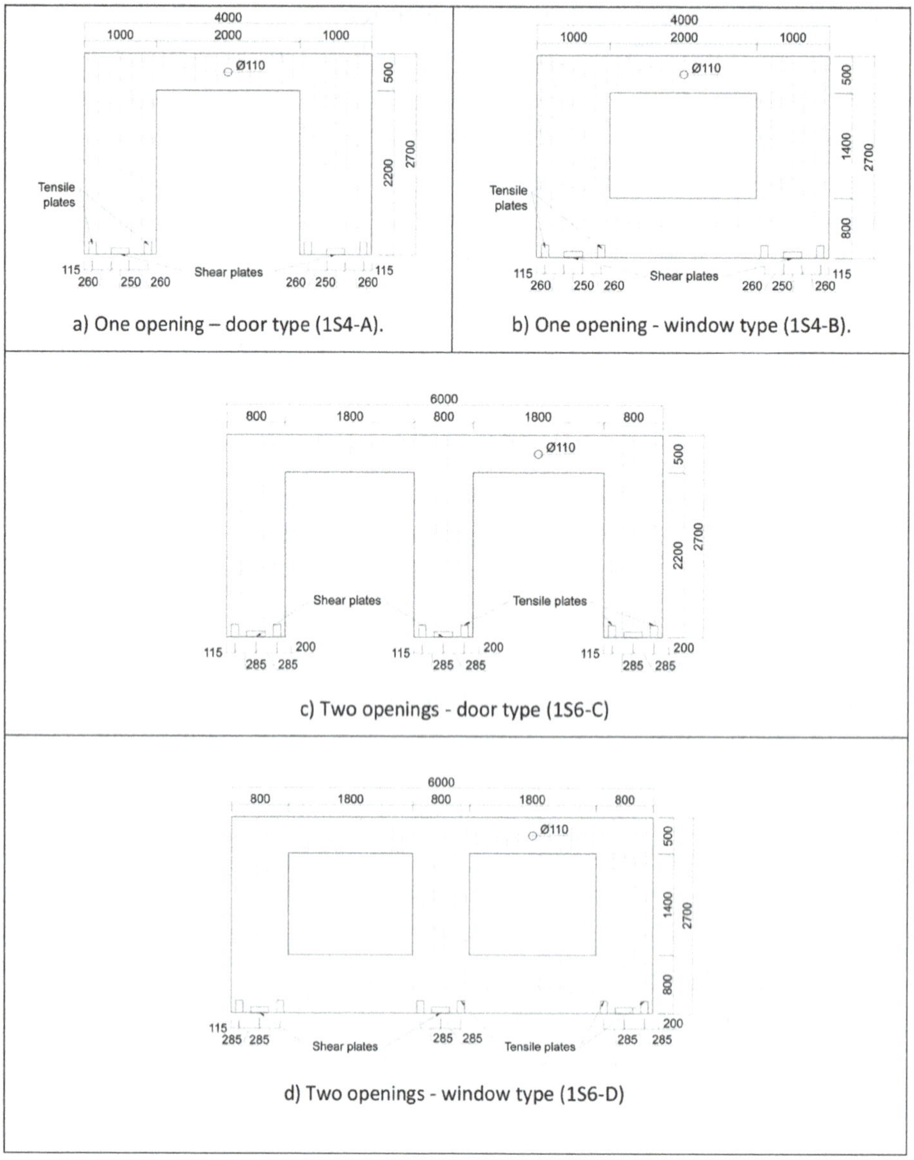

Figure 1. Configuration and dimensions in mm of the tested CLT panels. (a) One opening—door type (1S4-A). (b) One opening—window type (1S4-B). (c) Two openings—door type (1S6-C). (d) Two openings—window type (1S6-D).

(a) (b)

Figure 2. Two opening—door type (1S6-C): (a) overall view and (b) connection detail.

(a) (b)

Figure 3. Connections: (a) tensile-type and (b) shear type.

The specimens were bolted down to a steel beam foundation with a double-tee Japanese profile (H300 × B300 × WT10 × FT15). The lateral load was applied to the pin joint of the CLT panels at a height of 2.450 m using a pair of oil jacks (**Figure 4**). 'C' steel-loading beams were bolted to the top of the CLT wall through a steel plate of 300 × 600 × 10 mm with screws with a diameter of 7 mm. In addition, lateral guides with rollers were used to ensure a steady and consistent unidirectional movement of the walls. Cyclic loading was applied with three cycles at each target deformation angle: 1/450, 1/300, 1/200, 1/150, 1/100, 1/75, 1/50 and 1/30 rad.

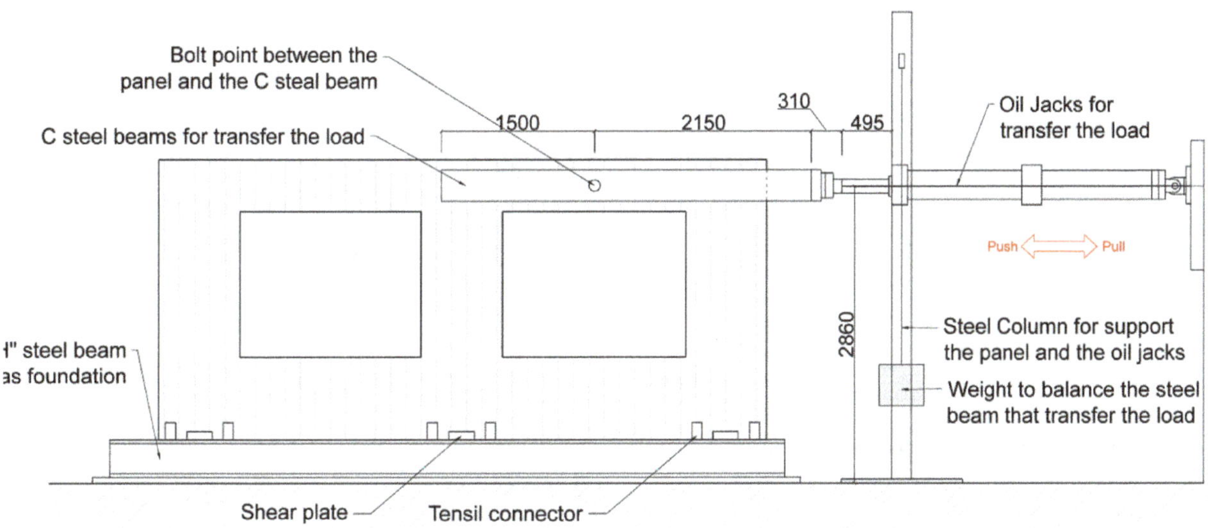

Figure 4. Experimental set-up (dimensions in mm).

Load was measured by the load cell set at the end of the oil jack, horizontal displacement of the panel at the height of loading points and near the bottom of the panel. Displacement transducers measured relative joint displacements (**Figure 5**). The axial strain of the bolt of each joint is measured using strain gauges.

Figure 5. Measure instruments: (a) strain gauges applied on the lateral wall of the specimen 1S4-B4; (b) displacement transducer in the lateral walls and (c) displacement transducer at the joint point between lateral walls and beams.

The transducers were placed at the bottom and at the top of the panel (at a height of 2450 mm). The one at the top (**Figure 6**) allowed measuring the displacement of the entire panel, whereas the three at the bottom measured relative displacement between the wall base and the steel foundation. In the graphs *P*-δ that will be shown to present results, *P* is the shear load transferred from the two actuators while δ is the displacement defined as the difference of the translation at the top of the panel and the translation at the bottom of the panel, calculated as the average value given by the displacement recorded from Chapters 2 to 4 (**Figure 6**).

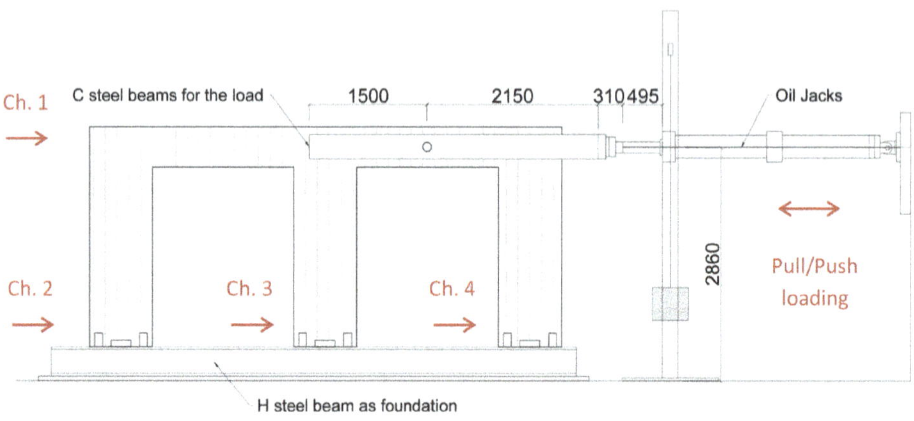

Figure 6. Measure points and channels (dimensions in mm).

2.2. Experimental tests results

2.2.1. *Timber panels response*

The cyclic tests results are reported and discussed in this paragraph. **Figure 7** summarizes the hysteresis curves obtained for each specimen specifying the cause of the failure and the crack position. **Table 2** reports the following elements:

- P_{max} is the maximum load derived from the load-deformation curve;
- δ_u is the ultimate deformation derived from the load-deformation curve;
- k_e is the value of the initial stiffness and is calculated as the slope of the line that connects the points of 0.2 P_{max} and 0.5 P_{max} on the hysteresis curve;
- P_y is the yield strength and is calculated as the value of the load at the point of intersection between the load-deformation curve and the 0.001-rad offset of the line corresponding to the initial value.

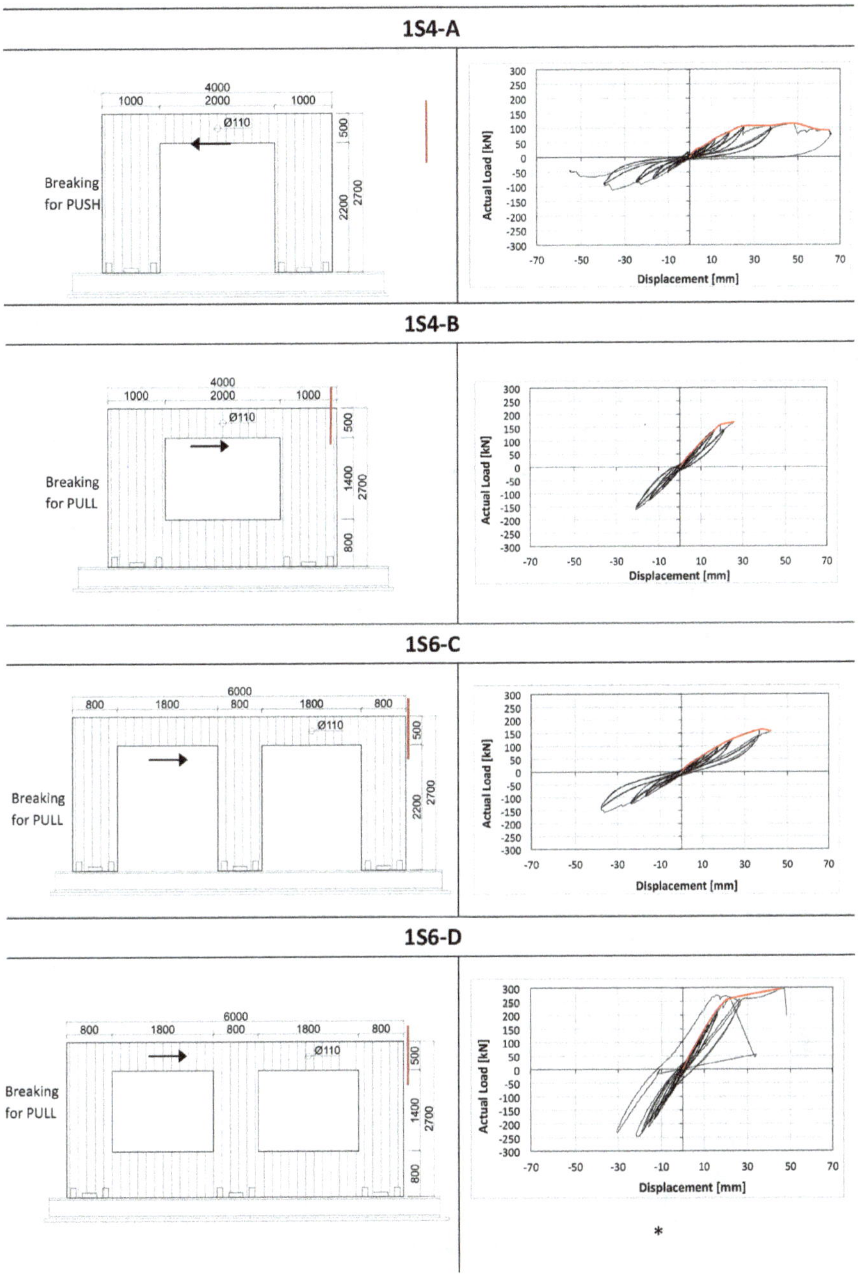

Figure 7. Test specimens and corresponding hysteresis curves (dimensions in mm). Backbone curve shown in red.

Specimen		P_{max} (kN)	δ_u (mm)	P_y (kN)	k_e (kN/mm)
One opening	1S4-A	112.43	65.89	103.40	4.21
	1S4-B	167.90	25.95	162.50	8.67
Two openings	1S6-C	160.60	42.15	149.61	4.99
	1S6-D	274.25	47.05	256.94	13.23

Table 2. Tabular results of the experimental tests (for the symbols meaning, see the body text).

For all the specimens, failure occurred at the upper corner of the opening and cracks always grew up from the corner of the opening to the top of the panel. The general behaviour was very brittle for all the panels with the exception of panel A for which a ductile behaviour can be recognized, entailing that the one-door-opening sample was also the one with the maximum dissipation of energy and the maximum deformation. The maximum strength was obtained for sample D but in this case the bending and sliding of the panel affected the data. Anyway, it can be observed that the maximum strength of the window type (panels B and D) is higher than that for the door type (panels A and C).

The initial stiffness k_e was calculated as the slope of the line that connects the points of 0.2 P_{max} and 0.5 P_{max} on the hysteresis curve. For calculating these values, the backbone curves of each hysteresis loop have been defined. The window types (panels B and D) showed a higher value of the initial stiffness k_e. Samples B and D are more resistant but allow lower deformations, whereas samples A and C have lower strength but higher deformations. Generally speaking, the backbone curve (**Figure 7**) represents the relationship between the maximum force and the corresponding deformation for each cycle of deformation angle to which the cyclic loading was performed. **Figure 8** shows the four backbone curves, allowing to compare the different behaviour of the wall-panel specimens.

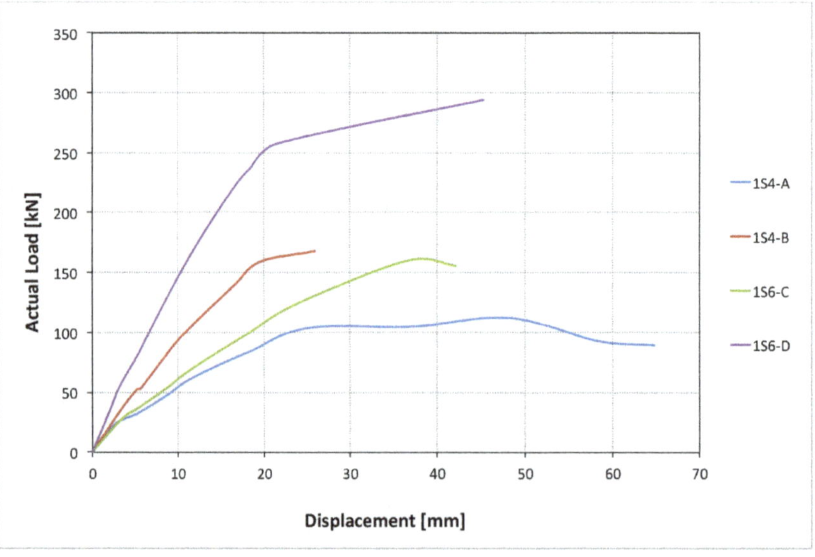

Figure 8. Comparison between the backbone curves of the four specimens.

The yielding strength was obtained as the point of intersection between the backbone curve and the line parallel to the initial strength and distant from this one of 2.45 mm in the graph. The value of 2.45 mm was calculated as 0.001 rad × 2450 mm, where 0.001 rad is given by the Japanese regulations and 2450 mm is the height of the top displacement transducer in the panel. The four wall panels have a prevalent brittle behaviour, except for specimen A that has a more ductile performance. Moreover, for all the four panels, it can be noted that the hysteretic loop has a typical pinched behaviour that is especially evident for panel A (**Figure 7**). In this case, in fact, the behaviour of the specimen, characterized by the presence of a wide open, was more affected by the behaviour of the connections than for other panels. As it will be discussed later, the screws used for panel A, after the test, were irreversibly damaged and their movement during the experiment caused the ovalisation of the holes. Thus, for panel A, during the model phase, the pinching effect, the degradation of stiffness and strength cannot be neglected.

2.2.2. Connections response

The configuration of the openings and the layout and design of joints strongly influenced the overall behaviour of a CLT structural system. For panels B, C and D, where the ratio between the openings area and the total area of the panel was lower than that for panel A, the overall response was ruled by the timber role. Moreover, for this panel the behaviour of the connectors largely affected the response of the specimen to lateral load, even if the failure is caused by the failure at the corner of the opening. It can be assessed that the connectors have basically all the same behaviour: they have a very stiff performance for the first cycles of the test, but, due to the wood crushing around the nails, their stiffness widely decreases during the experiment. The maximum strength is always around the same value, showing that although the connectors do not break, they all reach the yielding point. Moreover, their behaviour confirms that each connector has a negligible resistance to the tensile force that must be considered during the numerical model (Section 4) in order to correctly predict the rigid rotation and the stiffness of the specimens.

3. Analytical model

The load-bearing capacity and stiffness of fenestrated wood walls are mostly influenced by the size and layout of the openings. The reduction of strength observed during the experimental tests depends on the ratio of fenestrated area of the panel. Let us consider panels A and B, a wall without openings (index f). Let their ultimate strengths be, respectively, as P_A, P_B and P_f. One can suppose that the maximum strength of a panel with the same geometrical and boundary condition but with openings will be directly proportional to the value of P_f, namely:

$$P_A = P_f \times \alpha_A \tag{1}$$

$$P_B = P_f \times \alpha_B \tag{2}$$

where α_A and α_B are coefficients of <1. More in general this formula can be rewritten as:

$$P_n = P_f \times \alpha_n \tag{3}$$

For a generic panel with generic openings and boundary conditions, in order to find a bond between the strength and the geometrical values, the following considerations can be made:

- P_A and P_B will be directly proportional to the length $\Delta^{\underline{a}}$ of full-height wall segments and inversely proportional to the openings height;

From Eqs. (1) and (2), the ratio between P_A and P_B is equal to the ratio between α_A and α_B:

$$\frac{P_A}{P_B} = \frac{\alpha_A}{\alpha_B} \tag{4}$$

- The beams above and below the opening contribute in transferring load;
- All the panels have the same height and similar openings.

$$[P_A \cdot (h_A L_A - h_A b_A)] : b_A = [P_B \cdot (h_B L_B - h_B b_B)] : b_B \tag{5}$$

$$\frac{P_A}{P_B} = \frac{b_A}{b_B} \cdot \frac{h_B(L_B - b_B)}{h_A(L_A - b_A)} \tag{6}$$

Therefore, the following analytical formulation is proposed: where b and h are, respectively, the width and the height of the openings, while L is the width of the panel (**Figure 9**). P is the maximum shear strength attained by each panel during the experimental tests with the exception of panel D, whose value is not the one given by the experiment but the one corresponding to a deformation of 30 mm that, looking at the P-δ curve (**Figure 8**), and comparing its behaviour with the other three panels, seems more reliable.

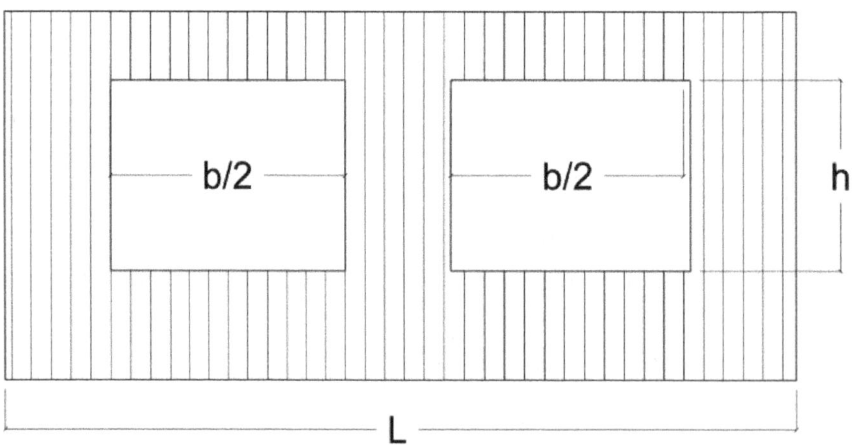

Figure 9. Symbols for the formulation of the ultimate strength in the analytical model.

The geometrical dimensions for the tested panels are reported in **Table 3**. **Table 4** shows the application of the proposed analytical expression for the four panels that have been tested in Tsukuba. The ratio between the ultimate strength of two panels determined during the experimental session is compared with the value obtained by comparing the geometrical conditions.

Specimen	*P* (kN)	*L* (m)	*b* (m)	*h* (m)	
One opening	1S4-A	112.42	4	2	2.2
	1S4-B	167.90	4	2	1.4
Two openings	1S6-C	160.64	6	3.6	2.2
	1S6-D	272.05	6	3.6	1.4

Table 3. Mechanical and geometric data for each tested panel.

$$\frac{P_A}{P_B} = 0.669 = \frac{b_A}{b_B} \cdot \frac{h_B(L_B - b_B)}{h_A(L_A - b_A)} = 0.636 e\% Error = 5.2\% \quad (8)$$

$$\frac{P_A}{P_C} = 0.699 = 0.667 \quad \% Error = 4.8\% \quad (9)$$

$$\frac{P_A}{P_D} = 0.413 = 0.424 \quad \% Error = 2.6\% \quad (10)$$

$$\frac{P_B}{P_C} = 1.045 = 1.048 \quad \% Error = 0.3\% \quad (11)$$

$$\frac{P_B}{P_D} = 0.617 = 0.667 \quad \% Error = 7.5\% \quad (12)$$

$$\frac{P_C}{P_D} = 0.590 = 0.6360\% Error = 7.2\% \quad (13)$$

Table 4. Application of the proposed formulation and errors between the analytical values and the experimental value.

The error given by the analytical value never exceeds 8%, showing therefore that the mathematical formulation can predict fairly closely the ultimate strength of panels with the same geometry, characteristics and boundary conditions. However, it must be noted that this relationship gives acceptable results when panels are similar. If configuration of the openings or dimensions of the compared panels such as height and thickness change and, for example, the openings are not symmetrical, the proposed equation is too simple and it will not lead to reliable results.

Moreover, no tests on no-fenestrated panels have been conducted, so it was not possible to compare the results with a common value P_f. Thus, the analytical formulation can be used only if the ultimate strength of one of two panels is already known and the panels have the same height. The mathematical model proposed in this paragraph is based on rough calculations and is therefore very approximate; however, it can be interpreted as a way to provide first information about the tendency of the reduction of racking strength of CLT shear walls with openings.

4. Finite element model

4.1. Description and mechanical parameters

The tested panels were modelled in SAP2000 by using a two-dimensional (2D) schematization with *"Shell-Layered/Nonlinear"* model [16]. The material properties adopted in the finite element model are listed in **Table 5**.

Modulus of elasticity—lower value (N/mm²)	E_{low}	**4200**
MOE—average value (N/mm²)	E_{av}	5200
Maximum bending strength (N/mm²)	σ_b	11,6
Modulus of elasticity—outer layers (N/mm²⁰)	$E_1 = E_3 = E_h$	173.33
Modulus of elasticity—inner layer (N/mm²)	$E_2 = E_v$	5200
Rolling shear modulus (N/mm²)	$G_{12} = G_{23}$	100
Longitudinal shear modulus (N/mm²)	G_{13}	400
Tensile strength—minimum value (N/mm²)	6.1.1. σ_t	12
Tensile strength—average value (N/mm²)	σ_t	16
Density (kg/m³)	ρ	439
Poisson's coefficients	ν	0.35

Table 5. Material properties adopted in the finite element model.

Under lateral loads, the connectors exhibit two different mechanisms of deformation. In the vertical direction, the anchors are subjected to tension, while in the horizontal one they experience shear deformation. These two deformation mechanisms are incorporated into the model by using individual springs for each of it, which act in unison. To find the stiffness and ultimate strength, tests on single-anchor elements should be conducted. In the present case, only the tensile connector (UT) has been previously subjected to monotonic load tests to correctly define its behaviour when subjected to tension.

The stiffness, strength and ductility of the steel connections are determined according to the Yasumura and Kawai procedure [17]. This procedure was initially proposed for the evaluation of wood-framed shear walls. The ultimate strength P_u is calculated so that the equivalence of the deformation energies is achieved by assuming an elasto-plastic load-displacement curve.

Figure 10 shows the definition of the bilinear curve that schematizes the behaviour of the tensile connectors. The contact—valid for tensile connectors—has been explicitly modelled using a set of compression-only springs identified at each point of the boundary mesh. For simulating the presence of the steel foundation, nodes with centre-to-centre distance of 10 cm have been generated at the base of the wall and all the degrees of freedom have been constrained. In the wall-to-floor contact, zero-length multilinear springs connect the nodes of the wall panel to the floor nodes. The compression-only springs are stiff in compression, and allow free movements away from it when subjected to tension. These springs are distributed along the contact between the wall and the floor. The friction between the steel beam foundation and the timber wall element is described by using spring elements with symmetrical and rigid-plastic behaviour placed along the whole length of the lower edge of the panel between the foundation nodes and the panel. The sliding resistance is described by the following equation:

$$F_f = k_f \bullet F_N \tag{7}$$

where F_N is the axial force at the current analysis step, k_f the static friction coefficient and F_f is the static friction force. The friction coefficient between the rough concrete and the CLT wooden surface was estimated as equal to 0.7 instead of the usual value of 0.4 used for two pieces of timber. In a proper schematization of the panel, the friction force should be calculated for each node taking into account the effective axial force that lies on each spring. Springs are stiff until the shear flow in the contact zone does not attain the estimated friction force. After this stage, friction springs have constant load-bearing capacity and resist sliding of panel in combination with non-linear springs that represent shear connectors.

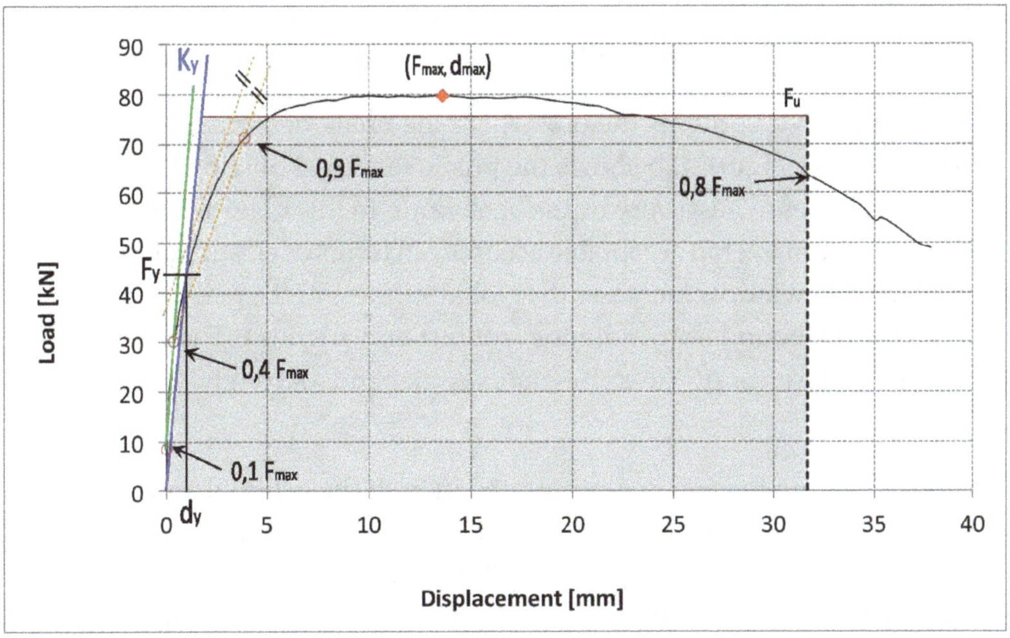

Figure 10. Definition of the bilinear curve (kN-mm) determined according to the *Yasumura and Kawai procedure* [17].

A pushover analysis was performed with a control of imposed displacement.

4.2. Results from finite element modelling

Confirming the experimental observation, the break occurs in the inner-cross layer due to the maximum tension attained in the corner of the opening of panel A (**Figure 11**). The maximum strength of 12 MPa is attained for a corresponding displacement of 17mm and 83-kN force.

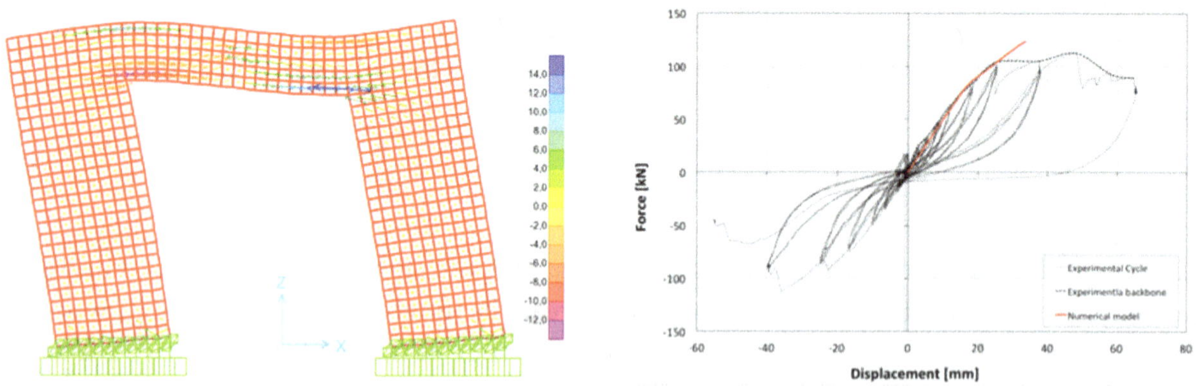

(a) (b)

Figure 11. Panel A: (a) maximum and minimum tension stresses in MPa at the last analysis step; (b) numerical push-over curve compared to the experimental curve.

The force-deformation response obtained matches quite good to the experimental response for the elastic behaviour. When the panel starts to break and the behaviour became plastic, the CLT shear wall is subjected to large displacement for small increments of load.

For panel B, the upper left corner is the one where the break occurred, as seen in the experimental test (**Figure 12a**). **Figure 12b** shows the pushover curve obtained for the shear wall B. Panel B, contrary to panel A, has a very brittle behaviour. In this case, the yielding point is near the breaking point and an overall acceptable accuracy in terms of elastic stiffness was obtained. The presence of the sub-window increases the global stiffness of the panel and highlights again the relevant role of the boundary conditions (contact and friction). The overall behaviour of panel C, due to the absence of the sub-windows, depends strongly from the UT and US connectors.

In this case, the maximum tension is concentrated in both the external and internal corners as shown in **Figure 13a**. Due to the eccentric position of the load joint (located not in the geometrical centre of the panel but in the centre of the right window), the maximum tension that brought to failure occurred in the inner corner. **Figure 13b** shows the comparison between the backbone curve and the pushover curve with the observation that the numerical model results approximate the experimental ones quite well.

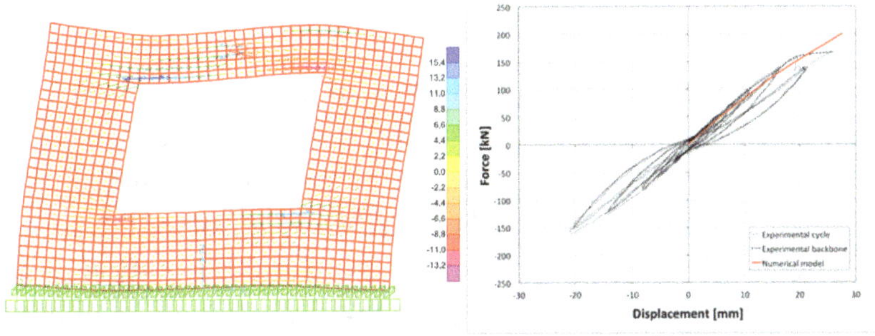

(a) (b)

Figure 12. Panel B: (a) maximum and minimum tension stresses in MPa at the last analysis step; (b) numerical push-over curve compared to the experimental curve.

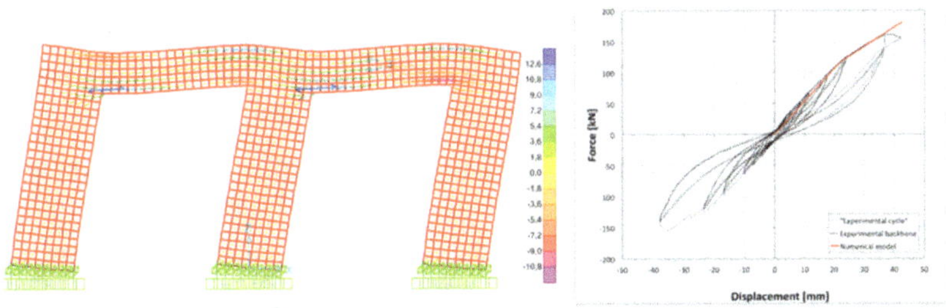

(a) (b)

Figure 13. Panel C: (a) maximum and minimum tension stresses in MPa at the last analysis step; (b) numerical push-over curve compared to the experimental curve.

As shown in **Figure 14a**, the stress concentration occurs in the corners of the windows and the breaking point corresponds with the inner corner of the left window confirming the experimental results. Also in this case, the sub-window contributes to increase the overall stiffness behaviour. In contrast with the other cases, for panel D, the pushover curve does not approximate exactly the stiffness of the panel (**Figure 14b**). The main reason of this result can be founded both in the general errors that occurred in the experimental session and in the general approximation of the boundary conditions. Other numerical analyses could be aimed at evaluating the energy dissipated by panels during cycles as done for masonry buildings [18].

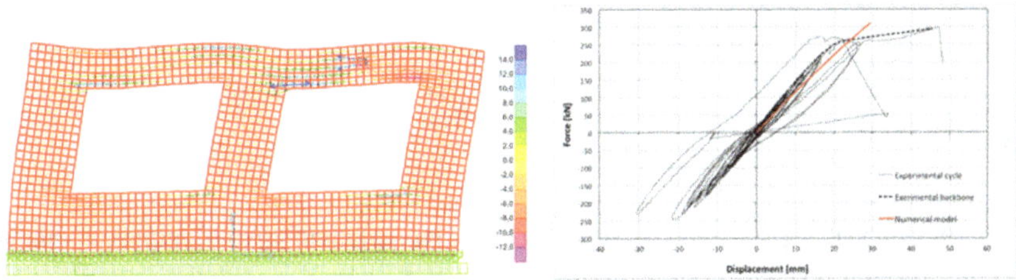

(a) (b)

Figure 14. Panel D: (a) maximum and minimum tension stresses in MPa at the last analysis step; (b) numerical push-over curve compared to the experimental curve.

5. Discussion and comparison between results

The main results obtained from experimental tests on CLT panels with openings have been compared and interpreted through analytical and numerical models. Concerning the experimental tests, failure occurred at the upper corner of the opening for all the specimens. The general behaviour was brittle for all the panels with the exception of the panel with a one-door opening, the most ductile and also the one with maximum dissipation of energy and deformation. The maximum strength was obtained for the sample with two windows but in this case the bending and sliding of the panel affected the results. Anyway, the maximum strength of the window type (panels B and D) was observed to be higher than that for the door type (panels A and C). An analytical model was adopted to predict the ultimate strength of panels similar to the tested ones, knowing the ultimate strength of one of two panels and the panels have the same height. The error given by the analytical method never exceeded 8%, showing therefore that the mathematical formulation can predict fairly closely the ultimate strength of panels with the same geometry, characteristics and boundary conditions. Finite element models confirmed, in terms of failure type and crack position, the experimental results. Moreover, the pushover curve obtained from the finite element procedure generally matched the experimental one quite well. Further analyses could be addressed to evaluate the out-of-plane resistance of the timber panels, by means of rocking analysis with proper boundary conditions, applying analogous concepts adopted for masonry panels [19, 20].

6. Conclusions

An experimental campaign aimed at evaluating the ultimate behaviour of CLT panels with openings was here described and interpreted with both analytical and numerical models. The four-wall panels were shown to exhibit a prevalent brittle behaviour, except for the specimen with one-door opening, more ductile. This response was reproduced quite well in the multi-layered finite element model. The position of the cracks at the ultimate limit state was correctly obtained from the numerical procedure, highlighting that the failure occurs at the corner of the openings, in different position depending on their size and configurations. The analytical model was capable to correctly evaluate the values of ultimate limit strength of walls with cut-out openings, with errors lower than 8%.

Acknowledgements

The research presented in this paper was funded by the Japanese company *'Nihon Sekkei System'* and supported by PRA2016 funding of the University of Pisa.

The authors would like to thank the Timber Structure Laboratory members (Department of Human and Social System, Institute of Industrial Science of the University of Tokyo) and Prof. Massimo Fragiacomo who provided technical expertise for the experimental testing.

Author details

Valeria Awad[1], Linda Giresini[1*], Mikio Koshihara[2], Mario Lucio Puppio[1] and Mauro Sassu[1]

*Address all correspondence to: linda.giresini@unipi.it

1 Department of Energy, Systems, Territory and Constructions Engineering (DESTEC), University of Pisa, Pisa, Italy

2 Department of Human and Social Systems, International Center for Urban Safety Engineering (ICUS), University of Tokyo, Tokyo, Japan

References

[1] Sassu, M., De Falco, A., Giresini, L., Puppio, M.L., Structural Solutions for Low-Cost Bamboo Frames: Experimental Tests and Constructive Assessments, Materials 2016, 2016, 9, 346; doi:10.3390/ma9050346.

[2] Handbook: cross-laminated timber (2011). Special Publication SP-528E, FPInnovations, edited by Gagnon S. and Pirvu C., QC, Canada.

[3] EN 1995-1-2 (Eurocode 5) (2004). Design of timber structures, Part 1-2: General—Structural fire design, CEN, Brussels, Belgium.

[4] Follesa, M., Christovasilis, I.P., Vassallo, D., Fragiacomo, M., and Ceccotti, A. Seismic design of multi-storey CLT buildings according to Eurocode 8. Ingegneria Sismica/International Journal of Earthquake Engineering, Special Issue on Timber Structures, Anno XXX-N.4 – Ottobre-Dicembre 2013.

[5] Studiengemeinschaft Holzeleimbau e.V. Building with cross laminated timber. Load-bearing solid wood components for walls, ceilings and roofs. Published: 04/2010, 2nd edition: 01/2010, 2011.

[6] Fragiacomo, M, Dujic, B, Sustersic, I. Elastic and ductile design of multi-storey crosslam wooden buildings under seismic actions. Engineering Structures 33, 2011, 3043–3053.

[7] Rinaldin, G., Amadio, G., Fragiacomo, M., A component approach for the hysteretic behaviour of connections in cross-laminated wooden structures. Earthquake Engineering & Structural Dynamics 2013; 42: 2023–2042.

[8] Bogensperger, T., Moosbrugger, T., Silly, G., *Verification of CLT-plates under loads in plane.* Proceedings of the 11th World Conference on Timber Engineering, Riva del Garda, Italy, 2010.

[9] Dujic, B., Klobcar, S., Zarnic, R., Influence of openings on shear capacity of wooden walls. Research report, University of Ljubljana and CBD Contemporary Building Design Ltd., Slovenia, 2005.

[10] Dujic, B., Aicher, S., Zarnic, R., Investigation on in-plane loaded wooden elements – influence of loading on boundary conditions. Otto Graf Journal, Materialprüfungsanstalt Universität. Otto-Graf-Institut, Stuttgart, Vol. 16, 2005.

[11] Dujic, B., Hristovsky, Zarnic R. *Experimental investigation of massive wooden wall panel system subject to seismic excitation*. Proceeding of the First European Conference on Earthquake Engineering. Geneva, Switzerland, 2006.

[12] Dujic, B., Klobcar, S., Zarnic, R., *Influence of Openings on Shear Capacity of Wooden Walls*. In: Proceedings of the 40th CIB-W18 Meeting, paper 40-15-6, Bled, Slovenia, 2007.

[13] Dujic, B., Klobcar, S., Zarnic, R., *Shear capacity of cross-laminated wooden walls*. Proceedings of the 10th World Conference on Timber Engineering, Miyazaki, Japan, 2008.

[14] Ceccotti, A., New Technologies for Construction of Medium-Rise Buildings in Seismic Regions: The XLAM Case, Structural Engineering International: Journal of the International Association for Bridge and Structural Engineering (IABSE), n. 18, pp. 156–165, 2008.

[15] Popovski, M. Lateral resistance of cross-laminated wood panels. In: Proceedings of the 11th world conference on timber engineering. Vol. 4. Trees e Timber Institute, pp. 3394–3403, 2010.

[16] Sap2000 V.14, CSI, Computers and Structures Inc., CA, USA.

[17] Yasumura, M, Kawai, N. Evaluation of wood framed shear walls subjected to lateral load. Meeting 30 of the Working Commission W18-Timber Structures, CIB. Vancouver, Canada, 1997, paper CIB-W18/30-15-4, 1997.

[18] Giresini, L., Energy-based method for identifying vulnerable macro-elements in historic masonry churches, Bulletin of Earthquake Engineering. 2006, Volume 14, Issue 3, pp 919–942. doi: 10.1007/s10518-015-9854-7.

[19] Giresini, L., Fragiacomo, M., Lourenço, P.B., Comparison between rocking analysis and kinematic analysis for the dynamic out-of-plane behavior of masonry walls. Earthquake Engineering and Structural Dynamics. 2015, Volume 44, Issue 13, pp. 2359–2376, doi: 10.1002/eqe.2592.

[20] Giresini, L., Fragiacomo, M., Sassu, M. Rocking analysis of masonry walls interacting with roofs. Engineering Structures, 2016, Volume 116, pp. 107–120. doi: 10.1016/j.engstruct.2016.02.041.

9

A Finite Element Method Model for Large Strains Analysis of Timber

Vincenzo De Luca

Abstract

In this report a Finite Element Method (FEM) model, within the continuum mechanics of solids, for mechanical long term response of timber structures is presented. The proposed model can analyze three-dimensional solids, within the theory for non-linear material orthotropic elastic-viscous-plastic. It can account ductile behaviour in compression and brittle behaviour in tension, under the kinematics hypothesis of large displacements and large strains. The work has been carried out with a general purpose FEM software code in which a specific stress-strain law for wood, by a proper subroutine, has been built in. The constitutive equations have been formulated by a multi-surface yield approach of viscoplasticity, each yield surface acts separately each other. This approach is specifically almost necessary for some problems, such as glued composite parts or steel bolt connections. Specialized solution algorithms, which adopt time-stepping and automating relaxation techniques, have been used to handle the behaviour of the loading path response for elastic-viscous-plastic or elastic-brittle behaviours. The model has been applied to examples to test the effectiveness of the suggested approach. The results obtained have shown the potentiality of the proposed model to effectively simulate the overall mechanical behaviour of timber.

Keywords: finite element method, timber model, elasto-visco-plastic model, multi-surface yield visco-plasticity

1. Introduction

Wood is one of the ancient materials employed in construction. It is commonly used as parts in buildings, such as bridges, roofs, sports halls, and floors. Today timber, especially in the glue-laminated technology, is frequently used in new constructions. Wood, as a material in the construction industry, has encountered a returned interest, mainly for its renewable nature, environmental compatibility and for its low energetic cost of production in comparison with concrete and steel.

Timber is advantageous, compared to other common construction materials, in wide spanning structures for its high ratio between load-carrying capacity and self-weight, along the longitudinal direction of grain. Timber structures have a good performance under dynamic loads for their elevated damping capacity. Although timber exhibits a consistent strength in the longitudinal direction for bending elements, in contrast it demonstrates a low value of tensile strength along the radial and tangential directions. This feature is unfavourable for connections, subjected to stress concentrations of tension, which can lead to crack development. However, timber is subjected to natural decay and its mechanical properties are sensitive to temperature and moisture.

An adequate mechanical description of timber material is an important task in scientific research on structural analysis of timber building. Also, an accurate stress analysis is particularly necessary for a composite system such as reinforced timber or pre-stressed timber, for the coupling and interaction of adhesive, lamination and reinforcements [1], and also for components used in connection joint [2].

In the scientific literature, many studies on modelling mechanics of wood can be retrieved [3]. However, continuum mechanics is the most common approach, being the basis of the finite element method (FEM) implementation. Within the theory of continuum mechanics, wood is assumed as an ideal homogeneous continuum [3], avoiding its complex nature. Wood is generally schematized as orthotropic elastic with three orthogonal material directions, corresponding to its longitudinal, radial and tangential growth fibre directions [3, 4]. However, experimental evidence [5] demonstrated that the extension of the Hooke's law to orthotropy can be acceptable only when the load intensity is low and not for high strain and for loading direction inclined with respect to the fibre orientation of the material [6].

Many theories on the wood material have been proposed in the literature. Among these, the earliest was that for wood in compression at an inclined angle of fibre, of Hankinson [7], which was extended by Goodman and Bodig [8] to a three dimension. Another model of wood was that of Tsai and Wu [9], which formulated a failure surface expressed as a scalar function of a polynomial tensor. Patton-Mallory et al. [2] used a three-dimensional model of wood, tri-linear elastic-plastic under compression stress and linear elastic, with a tension and shear cut-off, under tension stress.

Tabiei and Wu [10] proposed a continuum FEM model for timber which updated, in an incremental analysis, each component of the elastic moduli of the material stiffness matrix, on the basis of power functions, fitted to test data. This approach did not use any yield function of stress. Schmidt and Kaliske [11] proposed a three-dimensional model which adopted a multi-surface yield approach for elastic-plastic behaviour of wood. Recently, the adoption of FEM numerical codes to analyse the strength of a timber structural member has gained an increasing importance. Often the numerical codes are sophisticated and are specifically capable of representing the non-linear behaviour of the material [12–15] and some particular effects such as creep [16, 17].

The present work here discusses the following parts: the basic mathematical formulation of the continuum mechanics of solids according to the finite strain theory; the material model for timber: orthotropic, elastic-viscous-plastic, brittle in tension, ductile in compression, based on a multi-surface yield approach; the FEM formulation for large strains and large displacements with 3D solid hexahedral elements, numerical examples, results and discussion.

2. Materials and methods

The FEM is inherently developed according to the theory of the continuum mechanics of solids. In this work a general purpose FEM framework, with three-dimensional hexahedral elements and with a specific material model of wood, has been devised and software coded for the analysis of structural timber, which aims to reproduce the main mechanic performance of a timber element. In this code, the material model has been built in. In compression state, an elastic-viscous-plastic behaviour has been adopted, and in tension an elastic with brittle failure has been adopted. The stress tensor is accounted, at the integration points of an element, to compute the corresponding elastic-viscous-plastic material tangent modulus and the corresponding strain tensor. Failure is computed at a limit value of each component, in compression of both the stress tensor, representing yield, and strain tensor and in tension of the strain tensor only.

2.1. Introduction to the problem of continuum mechanics

According to the classical theory of the continuum mechanics of solids [18–20], the formulation of the FEM for finite strains is based on the description of the deformation state of a generic solid, subjected to a loading condition. The kinematic transformation of a body, from the state at the initial time 0, in the reference configuration ${}^{O}B$, to a successive state at the time t, in the current configuration ${}^{t}B$, is taken under consideration. ${}^{0}V$ and ${}^{t}V$ are the volumes of the body and ${}^{0}S$ and ${}^{t}S$ are the boundary surfaces at 0 time and t time, respectively. Displacements $\overline{\mathbf{u}}(\mathbf{x})$, on the portion ${}^{t}S_{\mathbf{u}}$ and tractions $\overline{\mathbf{t}}(\mathbf{x})$, on the portion ${}^{t}S_{\sigma}$ of the solid are specified as boundary conditions. Both body force loads, $\mathbf{b}$, and traction, $\mathbf{t}$, at the surface boundary, are assigned.

The solution to that problem consists to determinate the displacement of a point P, at a time t, of the body from the initial time to the current time, that is:

$$\mathbf{u} = {}^{t}\mathbf{x} - {}^{0}\mathbf{x} \tag{1}$$

where ${}^{0}\mathbf{x}$ and ${}^{t}\mathbf{x}$ are the coordinate vectors of the point P at the time 0 and t, respectively.

A basic tensor, which has an important role in the present theory, is the deformation gradient which can be calculated as:

$$\underline{\mathbf{X}} = \frac{\partial^t \mathbf{x}}{\partial^0 \mathbf{x}} \tag{2}$$

The quantities that govern the problem are distinguished if they are referred to the reference configuration, which is known, or to the current configuration, which is instead unknown.

The strains to which the body is subjected can be computed in the reference configuration by the Green-Lagrange strain tensor

$$\underline{\mathbf{E}} = \frac{1}{2}(\underline{\mathbf{X}}^{\mathrm{T}}\underline{\mathbf{X}} - \underline{\mathbf{I}}) \tag{3}$$

where $\underline{\mathbf{I}}$ is the unit tensor.

The stress tensor is computed through the constitutive law that expresses the stress tensor in terms of the strain tensor, by warranting that the stress and strain measures must be energetically conjugate. Accordingly to this, the Second Piola-Kirchhoff stress tensor, $\underline{\mathbf{S}}$, defined in the reference configuration, energetically conjugate to $\underline{\mathbf{E}}$, is calculated as:

$$\underline{\mathbf{S}} = \det(\underline{\mathbf{X}})\underline{\mathbf{X}}^{-1}\underline{\boldsymbol{\sigma}} \tag{4}$$

The true stress tensor or Cauchy stress tensor $\underline{\boldsymbol{\sigma}}$, is defined in the current configuration and it is related to $\underline{\mathbf{S}}$ as follows:

$$\underline{\boldsymbol{\sigma}} = \frac{1}{\det(\underline{\mathbf{X}})}\underline{\mathbf{X}^{\mathrm{T}}\mathbf{S}\mathbf{X}} \tag{5}$$

The measures of engineering stresses are expressed by the First Piola-Kirchhoff stress tensor, which is computed as

$$\underline{\mathbf{P}} = \underline{\mathbf{X}}^{\mathrm{T}}\underline{\mathbf{S}} \tag{6}$$

The above problem is ruled out by the equation of spatial momentum equilibrium in the reference configuration:

$$^{0}\rho\ddot{u} = div\,\underline{\mathbf{P}} - {}^{0}\rho\mathbf{b} \tag{7}$$

where $^{0}\rho$ is the mass density of the body, $\ddot{u}$ is the acceleration, which is set equal to $\ddot{u} = 0$ for the present static case. Also, the momentum equilibrium can be written by the weak form of the virtual work principle in the reference configuration:

$$\int_{^{0}V} \underline{\mathbf{P}} : grad\,\delta\mathbf{v}d^{0}V - \int_{^{0}V} {}^{0}\rho\mathbf{b}\delta\mathbf{v}d^{0}V - \int_{^{0}S_{\sigma}} \bar{t}\delta\mathbf{v}\Gamma d^{0}S = 0 \tag{8}$$

and in the current configuration:

$$\int_{{}^tV} \underline{\sigma} : \delta\underline{\mathbf{l}} d^tV - \int_{{}^tV} {}^t\rho\mathbf{b}\delta\mathbf{v}d^tV - \int_{{}^tS_\sigma} \bar{t}\delta\mathbf{v}d^tS = 0 \tag{9}$$

where Γ is the inverse Jacobian of surface and

$$\underline{\mathbf{l}} = \frac{\partial\mathbf{v}}{\partial^0\mathbf{x}}\frac{\partial^0\mathbf{x}}{\partial^t\mathbf{x}} = \frac{\partial}{\partial t}\left(\frac{\partial^t\mathbf{x}}{\partial^0\mathbf{x}}\right) : \frac{\partial^0\mathbf{x}}{\partial^t\mathbf{x}} = \underline{\dot{\mathbf{X}}}\,\underline{\mathbf{X}}^{-1} \tag{10}$$

is the spatial velocity gradient, $\delta\mathbf{v}$ is an admissible virtual velocity and $\delta\underline{\mathbf{l}}$ is the virtual velocity gradient that satisfy $\delta\mathbf{v} = 0$ on both ${}^0S_\mathbf{u}$ and ${}^tS_\mathbf{u}$.

By considering $d^tV = \det(\underline{\mathbf{X}})d^0V$, ${}^t\rho = \det(\underline{\mathbf{X}})^0\rho$, $d^tS = \Gamma d^0S$ and the spatial Kirchhoff stress tensor $\underline{\tau} = \det(\underline{\mathbf{X}})\underline{\sigma}$, we obtain the final expression of the principle of virtual work:

$$\int_{{}^0V} \underline{\tau} : \delta\underline{\mathbf{l}} d^0V - \int_{{}^0V} {}^0\rho\mathbf{b}\delta\mathbf{v}d^0V - \int_{{}^0S_\sigma} \bar{\mathbf{t}}\delta\mathbf{v}\Gamma d^0S = 0 \tag{11}$$

2.2. Finite element method framework

According to the FEM, the variables in the above equation can be discretized by interpolation forms, h_i and h_j, where $i, j = 1,..n$; n: number of nodes, and the integral equations are linearized to be integrated. The solutions of these non-linear equations can be obtained by the Newton-Raphson iterative method [18–20]. The FEM discretization gives the following system of equations for solving $d\mathbf{u}_j$:

$$\underline{\underline{\mathbf{K}}}d\mathbf{u}_j + \mathbf{R}_i - \mathbf{F}_i = 0 \quad \forall\{i\} : \mathbf{x}_i \notin S_u \tag{12}$$

$$\mathbf{u}_i = \bar{u}(\mathbf{x}_i) \quad \forall\{i\} : \mathbf{x}_i \in S_u \tag{13}$$

where the nodal force vector is:

$$\mathbf{F}_i = \int_{{}^0V} {}^0\rho\mathbf{b}h_i d^0V + \int_{{}^0S_\sigma} \bar{\mathbf{t}}h_i\Gamma d^0S \tag{14}$$

the nodal stiffness matrix is:

$$\underline{\underline{\mathbf{K}}} = \int_{{}^0V} \underline{\underline{\mathbf{D}}}\frac{\partial h_i}{\partial^t\mathbf{x}}\frac{\partial h_j}{\partial^t\mathbf{x}}d^0V - \int_{{}^0V} \underline{\tau}\frac{\partial h_i}{\partial^t\mathbf{x}}\frac{\partial h_j}{\partial^t\mathbf{x}}d^0V - \int_{{}^0S_\sigma} \bar{\mathbf{t}}h_i\frac{\partial\Gamma}{\partial\mathbf{u}_j}d^0S \tag{15}$$

the nodal residual vector is:

$$\mathbf{R}_i = \int_{{}^0V} \underline{\tau}\frac{\partial h_i}{\partial^t\mathbf{x}}d^0V \tag{16}$$

and the tangent stiffness matrix is:

$$\underline{\underline{\mathbf{D}}} = \frac{\partial \underline{\tau}}{\partial \underline{\mathbf{X}}} \underline{\mathbf{X}} \tag{17}$$

Also, to account that the orthotropic principal axes of the material can rotate with respect to the reference system, the material tensor, constitutive law, must be computed using a rotation tensor $\underline{\mathbf{R}}$ of the orthotropic axes in the reference system, which is determined from the spectral decomposition of the deformation tensor:

$$\underline{\underline{\mathbf{D}}} = \underline{\mathbf{R}}^{\mathrm{T}} \underline{\underline{\mathbf{D}}}_0 \underline{\mathbf{R}} \tag{18}$$

where $\underline{\underline{\mathbf{D}}}_0$ is the material tensor and $\underline{\underline{\mathbf{D}}}$ is the same material tensor rotated in the reference system.

The tangent stiffness matrix in general is non-symmetric and is dependent on the material tangent modulus. Its computation requires an adequate numerical approach which has to be generally relied to the constitutive relationships of the specific material.

The above non-linear system of Eqs. (12) and (13), at a global level, by applying the Newton-Raphson Method, is iteratively solved with the trial solution $\mathbf{u}_i$, updated with $d\mathbf{u}_j$. Coherently to solve the updated system, the above equations, the stiffness matrix, the residual vector and the force vector are updated. The convergence of the solution is checked on a measure of error, the root mean square of the $d\mathbf{u}_j$ or the nodal residual vector, the un-equilibrated force $-\mathbf{R}_i + \mathbf{F}_i$.

2.3. The constitutive material model

The non-linear behaviour of a material can be analysed by constitutive equations which relate stress to strain and other internal variables in a rate form. Within the finite strain hypothesis, constitutive equations must be formulated by warranting the principle of objectivity, that is, they must remain indifferent to the change of reference frame. This can be guaranteed by utilizing the objective tensor in constitutive equations.

In recent scientific literature some authors have proposed numerical solutions to the problem of continuum and consistent elastic-viscous-plasticity. Many scientific papers have been developed within the hypothesis of small strains [21–26], and others have focused their attention on finite strains hypothesis [27–31]. Many theoretical and numerical methods have been proposed in the literature to perform the procedure of stress update, in an incremental objectivity. The formulation of an incremental objective algorithm is based on constitutive equations, expressed as objective spatial rate, which are mapped to an intermediate configuration (in the present paper the relative variables are denoted with '^'). This configuration is a fictitious configuration which remains indifferent to rigid rotation. Also, the principle of objectivity is guaranteed by appropriate tensorial transformation of the body between spatial and material configuration.

Coherently to the above assertions, in the present work a convective or material representation [32, 33] has been employed. The problem is analysed by adopting the well-known multiplicative elasto-plasticity theory [32, 34] which postulates the multiplicative decomposition of the deformation gradient in an elastic part $\underline{\mathbf{X}}_e$ and a plastic (in the present work, the term 'plastic' means 'inelastic' in a more wide sense) part $\underline{\mathbf{X}}_p$:

$$\underline{\mathbf{X}} = \underline{\mathbf{X}}_e \underline{\mathbf{X}}_p \tag{19}$$

From the above discussion, the intermediate configuration can be described by the plastic part of the deformation gradient, $\underline{\mathbf{X}}_p$, except for a rigid rotation.

In order to formulate the elastic-viscous-plastic kinematics, a consistent transformation to the intermediate configuration is introduced for the spatial velocity gradient:

$$\underline{\mathbf{l}} = \dot{\underline{\mathbf{X}}}\,\underline{\mathbf{X}}^{-1} = \underline{\mathbf{l}}_e + \underline{\mathbf{l}}_p \tag{20}$$

where $\underline{\mathbf{l}}_e$ is the elastic part of the spatial velocity gradient and $\underline{\mathbf{l}}_p$ is the plastic part of the spatial velocity gradient.

The material velocity gradient is:

$$\underline{L} = \underline{\mathbf{X}}_e^{-1} \underline{\mathbf{l}} \underline{\mathbf{X}}_e = \underline{L}_e + \underline{L}_p \tag{21}$$

By mapping $\underline{\mathbf{l}}_e$ and $\underline{\mathbf{l}}_p$ to the intermediate configuration we obtain

$$\hat{\underline{L}}_e = \underline{\mathbf{X}}_e^{-1} \dot{\underline{\mathbf{X}}}_e \text{ the elastic part} \tag{22}$$

and

$$\hat{\underline{L}}_p = \dot{\underline{\mathbf{X}}}_p \underline{\mathbf{X}}_p^{-1} \text{ the plastic part} \tag{23}$$

of the velocity gradient $\hat{\underline{L}}$ in the intermediate configuration.

The stress power per unit reference volume is:

$$\mathcal{P} = \underline{\Sigma} : \hat{\underline{L}} \tag{24}$$

where the stress measure, work-conjugate to $\hat{\underline{L}}$, is the Mandel stress tensor $\underline{\Sigma} = \underline{\mathbf{C}}_e \hat{\underline{\mathbf{S}}}$, generally non-symmetric, $\underline{\mathbf{C}}_e = \underline{\mathbf{X}}_e^{\mathbf{T}} \underline{\mathbf{X}}_e$ is the elastic Cauchy-Green tensor, and $\hat{\underline{\mathbf{S}}} = \underline{\mathbf{X}}_p \underline{\mathbf{S}} \underline{\mathbf{X}}_p^{\mathbf{T}}$ is the Second Piola-Kirchhoff stress tensor in the intermediate configuration.

The constitutive relationships must conform to some restrictions [35], which derive from the second law of thermodynamics, expressed as the Clausius-Planck inequality. By considering an isothermal process and a purely mechanical case, the above inequality is expressed as plastic dissipation per unit reference volume:

$$D = \frac{1}{2}\underline{\hat{S}} : \underline{\dot{\hat{C}}}_e - \dot{\hat{\Psi}} \geq 0 \tag{25}$$

where

D is the dissipation; $\hat{\Psi} = \hat{\Psi}(\underline{\hat{C}}_e, \hat{\chi})$ is the free energy function, expressed in terms of $\underline{\hat{C}}_e$, and the internal variables $\hat{\chi}$ conjugate to the internal stresses $\hat{q}$.

The material time derivative of the free energy is

$$\dot{\hat{\Psi}} = \frac{\partial \hat{\Psi}}{\partial \underline{\hat{C}}_e} : \underline{\dot{\hat{C}}}_e + \frac{\partial \hat{\Psi}}{\partial \hat{\chi}} \cdot \dot{\hat{\chi}} \tag{26}$$

and the inequality results as:

$$D = \left(\underline{\hat{S}} - 2\frac{\partial \hat{\Psi}}{\partial \underline{\hat{C}}_e} \right) : \underline{\dot{\hat{C}}}_e + \underline{\Sigma} : \underline{\hat{L}}_p - \frac{\partial \hat{\Psi}}{\partial \hat{\chi}} \dot{\hat{\chi}} \geq 0 \tag{27}$$

The inequality (25) being valid for any admissible process of the material and from Eqs. (26) and (27) the constitutive equations are obtained as:

$$\underline{\hat{S}} = 2\frac{\partial \hat{\Psi}}{\partial \underline{\hat{C}}_e} \tag{28}$$

and

$$\hat{q} = \frac{\partial \hat{\Psi}}{\partial \hat{\chi}} \tag{29}$$

where $\hat{q}$ is the internal stress vector conjugate to the internal variables $\hat{\chi}$.

Hence the reduced dissipation inequality in a local form is:

$$D = \underline{\Sigma} : \underline{\hat{L}}_p - \hat{q} \cdot \dot{\hat{\chi}} \geq 0 \tag{30}$$

The hardening or softening law is expressed as follows:

$$\dot{\hat{\chi}} = \sum_m \dot{\gamma}_m \mathbf{H}_m(\underline{\hat{S}}, \hat{q}) \tag{31}$$

where γ_m are the plastic multipliers, $m = 1, \ldots, s$ and s is the number of multi-surfaces of yield, and $\mathbf{H}_m$ is the hardening or softening modulus, these quantities in the case of isotropic hardening or softening are scalars.

In the case of elastic response of stress, the response can be formulated in a non-rate form with a hyper-elastic potential in the intermediate configuration $\hat{\Psi}(\underline{\hat{C}}_e)$, this statement satisfy directly the principle of objectivity [28]. Thus

$$\hat{\Psi}(\hat{\underline{C}}_e) = \frac{1}{2}\hat{\underline{S}} : \hat{\underline{C}}_e \tag{32}$$

and [36]

$$\hat{\underline{S}} = 2\frac{\partial\hat{\Psi}}{\partial\hat{\underline{C}}_e} \tag{33}$$

is the Second Piola-Kirchhoff stress tensor in the intermediate configuration. From the differential of Eq. (32), the fourth-order elastic modulus tensor in the intermediate configuration can be obtained as:

$$\hat{\underline{\underline{D}}}_e = 2\frac{\partial\hat{\underline{S}}}{\partial\hat{\underline{C}}_e} = 4\frac{\partial^2\hat{\Psi}}{\partial\hat{\underline{C}}_e\partial\hat{\underline{C}}_e} \tag{34}$$

In the case of rate-independent elastic-plasticity, without the viscous effect, the point representative of stress cannot come outside the yield surfaces, identified by the yield functions $\hat{f}_m(\hat{\underline{S}}, \hat{q})$.

Note that in the following expression the stresses $\hat{\underline{S}}$ include the back stresses $-\hat{\underline{\beta}}$. The plastic multiplier γ_m can be computed by forcing the optimization conditions, named the Kuhn-Tucker loading/unloading conditions:

$$\dot{\gamma}_m \geq 0,\ \hat{f}_m \leq 0,\ \dot{\gamma}_m \hat{f}_m = 0 \tag{35}$$

and the consistency conditions:

$$\dot{\gamma}_m \dot{\hat{f}}_m = 0 \tag{36}$$

In the case of rate-dependent visco-plasticity, the actual stress can overcome out the yield surface, and consequently the above condition is no longer appropriate, but an overstress function $\hat{\Phi}_m(\hat{\underline{S}}, \hat{q})$ of the yield criteria can be defined such that:

$$\hat{\Phi}_m = 0\ when\ \hat{f}_m \leq 0 \text{ and } \hat{\Phi}_m = \hat{f}_m\ when\ \hat{f}_m > 0 \tag{37}$$

The viscous-plastic flow is evaluated by the following relaxation equation

$$\dot{\gamma}_m = \frac{1}{\eta}\hat{\Phi}_m \tag{38}$$

whose discretized form is:

$$\Delta\gamma_m = \frac{\Delta t_{t+\Delta t}}{\eta}\hat{\Phi}_m \tag{39}$$

where $\eta = 1$ is a viscosity parameter and $\eta \to 0$ in case of no viscid effects.

In order to determine the plastic stress response in Eq. (30), the plastic dissipation requires to compute the plastic part of the deformation gradient. From the principle of maximum plastic dissipation, the evolution equations of the inelastic strain tensors can be determined as normality rules:

$$\underline{\hat{L}}_p = \sum_m \dot{\gamma}_m \frac{\partial \hat{\Phi}_m}{\partial \underline{\Sigma}} \tag{40}$$

$$\dot{\hat{\chi}} = \sum_m \dot{\gamma}_m \frac{\partial \hat{\Phi}_m}{\partial \hat{q}} \tag{41}$$

The introduction of the consistency multipliers allows to distinguish the different material response upon each yield surface as follows:

$$\gamma_m = 0 \Rightarrow \textit{elastic response}$$

$$\gamma_m > 0 \Rightarrow \textit{plastic response.}$$

2.4. Objective integration algorithm

The constitutive equations can be approximately solved at a local level, with a general return mapping procedure. The time integration has been executed with a backward Euler scheme fully implicit in time, and an exponential map has been adopted for the plastic deformation gradient. According to this scheme of time integration, from a given state at the previous time step t, characterized by ${}^t\underline{\mathbf{X}}$ and ${}^t\underline{\mathbf{X}}_p$ and using the displacement increment $\Delta\mathbf{u}$, the deformation gradient ${}^{t+\Delta t}\underline{\mathbf{X}}$ can be determined. The trial elastic strain tensor $\underline{\hat{E}}^{trial}$ can be computed from the plastic deformation gradient ${}^t\underline{\mathbf{X}}_p$in the previous time step t.

The multiplicative decomposition, Eq. (19) is discretized

$${}^{t+\Delta t}\underline{\mathbf{X}} = {}^{t+\Delta t}\underline{\mathbf{X}}_e\, {}^{t+\Delta t}\underline{\mathbf{X}}_p \tag{42}$$

From the Eqs. (23) and (40) $\underline{\hat{L}}_p$ the flow rule can be recast in the following form:

$$\underline{\hat{L}}_p = \underline{\dot{\mathbf{X}}}_p \underline{\mathbf{X}}_p^{-1} = \sum_m \dot{\gamma}_m \underline{\hat{N}}_m \tag{43}$$

where

$$\underline{\hat{N}}_m = \frac{\partial \hat{\Phi}_m}{\partial \underline{\Sigma}} \tag{44}$$

is the viscous-plastic flow direction where the plastic potential coincide with the yield function, in an associative visco-plasticity.

The hardening or softening law has been discretized as follows:

$$^{t+\Delta t}\hat{\chi} = {}^{t}\hat{\chi} + \sum_{m} \Delta\gamma_m {}^{t+\Delta t}\mathbf{H}_m(\underline{\hat{\mathbf{S}}}, \hat{q}) \tag{45}$$

From the above Eq. (43), it is possible to obtain:

$$\underline{\dot{\mathbf{X}}}_p = \sum_{m} \dot{\gamma}_m \underline{\hat{\mathbf{N}}}_m \underline{\mathbf{X}}_p \tag{46}$$

the same relation can be integrated in time by an implicit backward Euler procedure, from t to $t + \Delta t$, with a time increment Δt and with an exponential shift [28]:

$$^{t+\Delta t}\underline{\mathbf{X}}_p = \exp\Big(\sum_{m} \Delta\gamma_m {}^{t+\Delta t}\underline{\hat{\mathbf{N}}}_m\Big) {}^{t}\underline{\mathbf{X}}_p \cong \Big(\underline{\mathbf{I}} + \sum_{m} \Delta\gamma_m {}^{t+\Delta t}\underline{\hat{\mathbf{N}}}_m\Big) {}^{t}\underline{\mathbf{X}}_p \tag{47}$$

where $\Delta\gamma_m = \Delta t \dot{\gamma}_m$ is the step increment of the consistent elastic-viscous-plastic multiplier. In the above discretization the direction of the plastic flow has been taken constant throughout the increment and is equal to its final value $^{t+\Delta t}\underline{\hat{\mathbf{N}}}_m$ which is assumed to be equal to that of the trial step.

Then the continuum elastic-viscous-plastic tangent stiffness matrix, which expresses the derivative of stress to respect to strain, it can be determined:

$$\underline{\underline{\hat{\mathbf{D}}}} = \left[\underline{\underline{\hat{\mathbf{D}}}}_e - \frac{\Big(\underline{\underline{\hat{\mathbf{D}}}}_e : \sum_{m} \hat{\mathbf{N}}_m\Big) \otimes \Big(\underline{\underline{\hat{\mathbf{D}}}}_e : \sum_{m} \hat{\mathbf{N}}_m\Big)}{\Big(\sum_{m} \hat{\mathbf{N}}_m : \underline{\underline{\hat{\mathbf{D}}}}_e : \sum_{m} \hat{\mathbf{N}}_m - \sum_{m} \frac{\partial \hat{\Phi}_m}{\partial \hat{q}} \cdot \mathbf{H}_m + \eta \cdot c_{creep}\Big)} \right] \tag{48}$$

where the creep term is expressed as follows:

$$c_{creep} = \left(\frac{1}{a_1 \sum_{m} \Delta\gamma_m}\right) \left(\frac{\sum_{m} \Delta\gamma_m}{\Delta t \dot{\overline{\varepsilon}}_{c0} t^{a_2}}\right)^{\frac{1}{a_1}} \tag{49}$$

where the power law for creep is evaluated [16] as:

$$\overline{\varepsilon}_c = \Delta t a_0 \overline{\boldsymbol{\sigma}}^{a_1} t^{a_2} \tag{50}$$

where

$\overline{\varepsilon}_c$ is the uniaxial creep strain;

$\dot{\overline{\varepsilon}}_{c0}$ is a reference creep strain rate; $\overline{\boldsymbol{\sigma}}$ is the equivalent stress; a_0 is the coefficient in the creep power law, which is strain rate; a_1: is the exponent of the equivalent stress in creep power law; a_2 is the exponent of time in creep power law; Δt: is the time increment.

The trial stress predictor is computed as

$$\underline{\hat{S}}^{trial} = \underline{\underline{\hat{D}}}_e : \underline{\hat{E}}^{trial} \tag{51}$$

and in discretized form

$$^{t+\Delta t}\underline{\hat{S}} = \underline{\underline{\hat{D}}}_e : {}^{t+\Delta t}\underline{\hat{E}}^{trial} \tag{52}$$

where

$$\underline{\hat{E}}^{trial} = \frac{1}{2}({}^{t+\Delta t}\underline{\mathbf{X}}^{\mathbf{T}\,t+\Delta t}\underline{\mathbf{X}} - \underline{\mathbf{I}}) \tag{53}$$

Substituting the Eqs. (42) and (47) into the elastic Green Lagrange strain tensor, we have:

$$\underline{\hat{E}} = \underline{\mathbf{X}}_p^{\mathbf{T}-1}\underline{\hat{E}}^{trial}\underline{\mathbf{X}}_p^{-1} \tag{54}$$

For all the multi-surface yields, by using the proper yield criterion, the trial overstress function $\hat{\Phi}_m^{trial}$ can be computed and hence the flow direction $\underline{\underline{\hat{N}}}_m^{trial}$ can be evaluated as norm of the deviatoric stress tensor $\| DEV(\underline{\hat{S}}^{trial}) \|$. In the case of elastic response, the values of the variables at current time $t + \Delta t$ are equal to their respective predictors. Instead, in the case of plastic response by using the exponential mapping (47), the stress point is mapped back to the yield surface.

Then, the increment of each plastic multiplier $\Delta\Delta\gamma_m$, discretized in time, in the case of plastic loading $^{t+\Delta t}\,\hat{\Phi}_m = 0$, can be solved for each $^{t+\Delta t}\Delta\gamma_m$ by means of a local iterative Newton-Rapshon Method.

For all the multi-surfaces, at the end of each local iteration, the intermediate configuration, described by $^{t+\Delta t}\,\underline{\mathbf{X}}_p$ and the internal variables are updated.

The updated stress tensor is evaluated as follows:

$$\underline{\hat{S}} = \underline{\underline{\hat{D}}} : \underline{\hat{E}} - \sum_m \Delta\gamma_m \underline{\hat{N}}_m \tag{55}$$

In the following expressions

$$^{t+\Delta t}\underline{\hat{E}}_p = {}^{t}\underline{\hat{E}}_p + {}^{t+\Delta t}\Delta\underline{\hat{E}}_p \tag{56}$$

$$^{t+\Delta t}\hat{\chi} = {}^{t}\hat{\chi} + {}^{t+\Delta t}\Delta\hat{\chi} \tag{57}$$

the increments are evaluated as follows:

$$ {}^{t+\Delta t}\Delta\underline{\hat{E}}_p = \sum_m {}^{t+\Delta t}\Delta\gamma_m \frac{\partial^{t+\Delta t}\hat{\Phi}_m}{\partial\underline{\hat{S}}} \tag{58} $$

$$ {}^{t+\Delta t}\Delta\hat{\chi} = -\sum_m {}^{t+\Delta t}\Delta\gamma_m \frac{\partial^{t+\Delta t}\hat{\Phi}_m}{\partial\hat{q}} \tag{59} $$

by introducing the residual vectors:

$$ {}^{t+\Delta t}\mathbf{r}_p = -{}^{t+\Delta t}\underline{\hat{E}}_p + {}^{t}\underline{\hat{E}}_p + {}^{t+\Delta t}\Delta\underline{\hat{E}}_p = 0 \tag{60} $$

$$ {}^{t+\Delta t}\mathbf{r}_\chi = {}^{t+\Delta t}\hat{\chi} - {}^{t}\hat{\chi} - {}^{t+\Delta t}\Delta\hat{\chi} = 0 \tag{61} $$

$$ {}^{t+\Delta t}r_\Phi = {}^{t+\Delta t}\hat{\Phi}_m({}^{t+\Delta t}\underline{\hat{S}}, {}^{t+\Delta t}\hat{q}) = 0 \tag{62} $$

which are solved for the variables:

$$ \left\{ \begin{array}{c} {}^{t+\Delta t}\underline{\hat{S}} \\ {}^{t+\Delta t}\hat{q} \\ {}^{t+\Delta t}\Delta\gamma_m \end{array} \right\} \tag{63} $$

By linearizing [37] the residual vectors, a system of linear equations is obtained:

$$ {}^{t+\Delta t}\mathbf{r}_p + \frac{\partial^{t+\Delta t}\mathbf{r}_p}{\partial\underline{\hat{S}}} : \Delta\underline{\hat{S}} + \frac{\partial^{t+\Delta t}\mathbf{r}_p}{\partial\hat{q}} : \Delta\hat{q} + \frac{\partial^{t+\Delta t}\mathbf{r}_p}{\partial\Delta\gamma_m} : \Delta\Delta\gamma_m = 0 \tag{64} $$

$$ {}^{t+\Delta t}\mathbf{r}_\chi + \frac{\partial^{t+\Delta t}\mathbf{r}_\chi}{\partial\underline{\hat{S}}} : \Delta\underline{\hat{S}} + \frac{\partial^{t+\Delta t}\mathbf{r}_\chi}{\partial\hat{q}} : \Delta\hat{q} + \frac{\partial^{t+\Delta t}\mathbf{r}_\chi}{\partial\Delta\gamma_m} : \Delta\Delta\gamma_m = 0 \tag{65} $$

$$ {}^{t+\Delta t}r_\Phi + \frac{\partial^{t+\Delta t}r_\Phi}{\partial\underline{\hat{S}}} : \Delta\underline{\hat{S}} + \frac{\partial^{t+\Delta t}r_\Phi}{\partial\hat{q}} : \Delta\hat{q} + \frac{\partial^{t+\Delta t}r_\Phi}{\partial\Delta\gamma_m} : \Delta\Delta\gamma_m = 0 \tag{66} $$

Dropping the superscript $t + \Delta t$ and writing:

$$ \underline{\underline{\Xi}}^{-1} = \underline{\underline{\hat{D}}}_e^{-1} + \Delta\gamma_m \frac{\partial^2\hat{\Phi}_m}{\partial\underline{\hat{S}}\partial\underline{\hat{S}}} \tag{67} $$

and

$$ \underline{\underline{\Omega}}^{-1} = \underline{\underline{H}}^{-1} + \Delta\gamma_m \frac{\partial^2\hat{\Phi}_m}{\partial\hat{q}\partial\hat{q}} \tag{68} $$

the solution can be obtained as

$$\begin{bmatrix} {}^{t+\Delta t}\Delta\underline{\hat{S}} \\ {}^{t+\Delta t}\Delta\hat{q} \\ {}^{t+\Delta t}\Delta\Delta\gamma_m \end{bmatrix} = \begin{bmatrix} \underline{\underline{\Xi}}^{-1} & \Delta\gamma_m \dfrac{\partial^2\hat{\Phi}_m}{\partial\hat{q}\partial\underline{\hat{S}}} & \dfrac{\partial\hat{\Phi}_m}{\partial\underline{\hat{S}}} \\ \Delta\gamma_m \dfrac{\partial^2\hat{\Phi}_m}{\partial\underline{\hat{S}}\partial\hat{q}} & \underline{\underline{\Omega}}^{-1} & \dfrac{\partial\hat{\Phi}_m}{\partial\hat{q}} \\ \dfrac{\partial\hat{\Phi}_m}{\partial\underline{\hat{S}}} & \dfrac{\partial\hat{\Phi}_m}{\partial\hat{q}} & 0 \end{bmatrix}^{-1} \cdot \begin{bmatrix} -\mathbf{r}_p \\ -\mathbf{r}_\chi \\ -r_\Phi \end{bmatrix} \tag{69}$$

where the consistent Jacobian, **J**, of the residuals is:

$$\frac{d\mathbf{r}}{d\mathbf{z}} = \mathbf{J}^{-1} = \begin{bmatrix} \underline{\underline{\Xi}}^{-1} & \Delta\gamma_m \dfrac{\partial^2\hat{\Phi}_m}{\partial\hat{q}\partial\underline{\hat{S}}} & \dfrac{\partial\hat{\Phi}_m}{\partial\underline{\hat{S}}} \\ \Delta\gamma_m \dfrac{\partial^2\hat{\Phi}_m}{\partial\underline{\hat{S}}\partial\hat{q}} & \underline{\underline{\Omega}}^{-1} & \dfrac{\partial\hat{\Phi}_m}{\partial\hat{q}} \\ \dfrac{\partial\hat{\Phi}_m}{\partial\underline{\hat{S}}} & \dfrac{\partial\hat{\Phi}_m}{\partial\hat{q}} & 0 \end{bmatrix}^{-1} \tag{70}$$

the vector of unknowns is:

$$\mathbf{z} = \begin{bmatrix} {}^{t+\Delta t}\Delta\underline{\hat{S}} \\ {}^{t+\Delta t}\Delta\hat{q} \\ {}^{t+\Delta t}\Delta\Delta\gamma_m \end{bmatrix} \tag{71}$$

and the vector of residual is:

$$\mathbf{r} = \begin{Bmatrix} \mathbf{r}_p \\ \mathbf{r}_\chi \\ r_\Phi \end{Bmatrix} \tag{72}$$

The system of equations is solved by the Newton-Raphson following iteration scheme:

$$\mathbf{z}_{k+1} = \mathbf{z}_k - \mathbf{J}^{-1}\mathbf{r}_k \tag{73}$$

where k is the iteration number. However, in a more advantageous mode the system of equations has been solved by partitioning Eq. (69) into submatrices.

At the start of the iteration loop the following positions are made:

$${}^{t+\Delta t}\underline{\hat{E}}_p^{(0)} = {}^{t}\underline{\hat{E}}_p \tag{74}$$

$${}^{t+\Delta t}\hat{\chi}^{(0)} = {}^{t}\underline{\hat{\chi}} \tag{75}$$

$${}^{t+\Delta t}\Delta\gamma_m^{(0)} = {}^{t}\Delta\gamma_m \tag{76}$$

At each iteration k the following quantities are computed:

$$^{t+\Delta t}\Delta\underline{\hat{E}}_p^{(k)} = -\underline{\underline{\hat{D}}}_e^{-1\,t+\Delta t}\Delta\underline{\hat{S}} \tag{77}$$

$$^{t+\Delta t}\Delta\hat{\chi}^{(k)} = \underline{\underline{\mathbf{H}}}^{-1\,t+\Delta t}\Delta\hat{q} \tag{78}$$

$$^{t+\Delta t}\Delta\gamma_m^{(k)} = {}^{t+\Delta t}\Delta\gamma_m \tag{79}$$

hence the internal variables are updated as follows:

$$^{t+\Delta t}\underline{\hat{E}}_p^{(k+1)} = {}^{t+\Delta t}\underline{\hat{E}}_p^{(k)} + {}^{t+\Delta t}\Delta\underline{\ddot{E}}_p^{(k)} \tag{80}$$

$$^{t+\Delta t}\hat{\chi}^{(k+1)} = {}^{t+\Delta t}\hat{\chi}^{(k)} + {}^{t+\Delta t}\Delta\hat{\chi}^{(k)} \tag{81}$$

$$^{t+\Delta t}\Delta\gamma_m^{(k+1)} = {}^{t+\Delta t}\Delta\gamma_m^{(k)} + {}^{t+\Delta t}\Delta\Delta\gamma_m^{(k)} \tag{82}$$

The next step begins by computing the stress tensor and the internal stress tensor. The iterations progress until the norm of the residual vectors stays behind a prefixed tolerance.

At this point each overstress function can be defined as follows:

$$\hat{\Phi}_m(\underline{\hat{S}},\hat{q}) = |DEV(\hat{S}_{ij}) + \hat{q}_{ij}| - \frac{3}{2}s_{mat} - s_{flow} \tag{83}$$

where:

i,j are the indexes which define the nine components of the stress and strain tensors; DEV is the deviatoric operator; $s_{mat} = [\underline{\hat{N}}_m \cdot (\underline{\underline{\Xi}} : \underline{\hat{N}}_m) + \underline{\hat{N}}_m \cdot (\underline{\underline{\Omega}} : \underline{\hat{N}}_m)]\Delta\gamma_m + [\mathrm{r} \cdot (\underline{\underline{\Xi}} : \underline{\hat{N}}_m) + \mathrm{r} \cdot (\underline{\underline{\Omega}} : \underline{\hat{N}}_m)]$; $s_{flow} = Y_m\left(1 + \frac{\overline{\varepsilon}_p + \Delta\overline{\varepsilon}_p}{\overline{\varepsilon}_0}\right) \cdot \left(\frac{\Delta\overline{\varepsilon}_p}{\Delta t \cdot \dot{\overline{\varepsilon}}_{c0} t^{a_2}}\right)^{\frac{1}{a_1}}$ is the stress flow due to creep effects; $s_{flow} = Y_m \left(1 + \frac{\overline{\varepsilon}_p + \Delta\overline{\varepsilon}_p}{\overline{\varepsilon}_0}\right)$ is the stress flow without creep effects; $\overline{\varepsilon}_p$is the equivalent accumulated plastic strain; $\Delta\overline{\varepsilon}_p$is the increment of the equivalent plastic strain; $\overline{\varepsilon}_0$is a reference strain assumed as the strain correspondent to Y_m; $\dot{\overline{\varepsilon}}_{c0}$ is a reference creep strain rate; m are multi-surface indexes; $\hat{S}_{ij}$ are the components of the stress tensor $\underline{\hat{S}}$; $\hat{q}_{ij}$ are the components of the hardening tensor $\hat{q}$; Y_m is the initial yield value of the m−th yield surface which is assumed as follows

$$Y_m = \begin{cases} Y_{ij}^t & if\ \hat{S}_{ij} \geq 0 \\ Y_{ij}^c & if\ \hat{S}_{ij} < 0 \end{cases} for\ i,j = 1,\ldots,3 \tag{84}$$

Y_{ij}^t are tensile values and Y_{ij}^c are compressive values, nine components which reduce to six for symmetry. The initial yield values are material parameters, which can be identified by experimental tests.

To enhance the convergence in the equilibrium iteration, with the Newton-Raphson Method, the consistent elastic-viscous-plastic tangent stiffness matrix can be used instead of the continuum elastic-viscous-plastic tangent stiffness matrix (48). This is the consequence of the discretization of the consistency condition in correspondence of the rate-independent elastic limit. It can be obtained by substituting the matrix $\hat{\underline{\underline{D}}}_e$, in the expression (48) the following pseudoelastic material stiffness matrix:

$$\underline{\underline{\mathbf{P}}} = \hat{\underline{\underline{D}}}_e^{-1} + \sum_m \Delta\gamma_m \frac{\partial^2 \hat{\Phi}_m}{\partial \hat{\underline{S}} \partial \hat{\underline{S}}} \tag{85}$$

2.5. The finite element method implementation

The previous non-linear constitutive material model has been implemented in a three-dimensional finite element code, based on the displacement approach. The geometry was schematized by adopting linear hexahedral solid iso-parametric elements with eight node. The spatial integration was carried out for each element with an 8-point Gauss quadrature rule.

Within the framework of the Newton-Raphson iteration method, the solution of the non-linear equilibrium equation system was found for the displacements as unknown variables.

The procedure implemented in the software code is listed in **Table 1**.

---

1. Read input file with the following data:
   a. Material constants
   b. Failure constants
   c. Point coordinates of finite elements
   d. Volume vertex points
   e. Groups of element materials
   f. Boundary points restrained
   g. Time-load increment functions
   h. Time steps at each time-load increment
   i. Force-loads
   j. Mass-proportional loads
   k. Volume element subject to mass-proportional loads
   l. Points subject to force-loads
2. Apply the load increment at a prescribed time-step
3. Solve the system of non-linear equations applying the Newton-Raphson Method (NRM) at a global level:
   a. While (Error>tolerance) and (Residue>tolerance) do
   b. Begin
   c. Execute an internal NRM loop for the elastic-viscous-plastic elements
   d. Compute the element stiffness and assemble the global stiffness matrix
   e. Compute the element force and assemble the global force vector
   f. Compute the global residual vector
   g. Update the global correction

---

h. Update the restrains

i. Update the assigned displacements

j. Solve the system of equations

k. Update the global displacements

l. Compute displacement Error and force Residue for check convergence

m. Check against the convergence criteria. The procedure goes to step 3 when the convergence is not reached, and solve the global system of non-linear equations on the vector of residual forces. On the contrary, the procedure progress to a successive time-step, go to step 2 by applying a load increment, when the convergence is reached.

n. End of loop

4. Update the global state, stress tensor, strain tensor and accumulated plastic strain

5. Update the total displacement vector

6. Update the geometrical nonlinearity

7. Output the results, displacements, stresses, strains and accumulated plastic strain.

Table 1. Scheme of the implementation of the software code.

The FEM code was employed to simulate the load-displacement behaviour of white spruce timber up to and beyond failure. Simple tests in tension and in compression have been run to check the ability of the computational code.

2.6. Comparison with experimental data and numerical example

As a check to the suggested implementation, application examples by using the present model have been simulated with the software code. The hexahedron 3D solid elements have been used to mesh timber. Note that the purpose of these tests is to make a preliminary check of the proposed model. Subsequently, the improvement of this model and of its parameters, and further checks against experimental data, is a future task.

To verify the present model preliminarily, a comparison between numerical results and experimental data has been reported here. The experimental data concern a test on a specimen of white spruce under compressive loading, with the loading axes set along the longitudinal fibre of wood. The specimen has been dimensioned with size 20 mm × 20 mm × 20 mm. The experimental data have been assumed from a previous work [6]. In this work, the specimen has been tested by an electro-mechanical press, whose computer controlled the displacement rate at 8 mm/min. The specimen has been loaded until it failed. In this test the elastic and post-elastic stress-strain path of the material has been data-logged. The numerical results have been obtained by referring to the dimensions and mechanical characteristics of the same specimen in order to reproduce the test on the real specimen effectively. The mechanical characteristics of the white spruce wood, assumed from Refs. [6, 38] in the present work, are reported in **Tables 2** and **3**. However, in this numerical test not any fracture has been considered, by reducing the compressive maximum strains respect to those measured in [6]. The creep parameters assumed from [16] are reported in **Table 4**. The simulation of the non-linear behaviour of a wood specimen under compression test has been executed by using a finite element mesh of eight node hexahedral elements, with 1000 elements and 1331 nodes. The specimen has been cinematically restrained at its bottom surface edge, whereas it has been subjected to an impressed displacement path at its top surface edge. The loading simulation has been carried out in control of displacement.

Property	E_{11}	G_{12}	G_{13}	G_{21}	E_{22}	G_{23}	G_{31}	G_{32}	E_{33}
	N/mm^2	N/mm^2	N/mm^2	N/mm^2	N/mm^2	N/mm^2	N/mm^2	N/mm^2	N/mm^2
Elastic modulus [6]	557	38	790	(1)	1010	829	(1)	(1)	12961
		ν_{12}	ν_{13}	ν_{21}		ν_{23}	ν_{31}	ν_{32}	
Poisson's ratio [38]		0.435	0.467	(2)		0.372	(2)	(2)	

(1) To satisfy the objectivity condition, the shear moduli, according to Lekhnitskii [39] are assumed as: $G_{ij} = \frac{E_{ii}E_{jj}}{E_{ii}(1+\upsilon_{ij})+E_{jj}(1+\upsilon_{ji})}$.

(2) The Poisson's ratios are taken as constant, and to satisfy the symmetry of the elasticity tensor, obeying the relations: $\frac{\upsilon_{ij}}{E_{ii}} = \frac{\upsilon_{ji}}{E_{jj}}$.

Table 2. Mechanical properties assumed for spruce wood in the numerical tests (E is the elastic modulus, G is the shear modulus and ν is the Poisson ratio, i, j = 1, 2, 3 are the indices, which correspond to longitudinal, radial and tangential fibre directions, respectively).

	Y^t_{11}	Y^t_{12}	Y^t_{13}	Y^t_{21}	Y^t_{22}	Y^t_{23}	Y^t_{31}	Y^t_{32}	Y^t_{33}
	N/mm^2	N/mm^2	N/mm^2	N/mm^2	N/mm^2	N/mm^2	N/mm^2	N/mm^2	N/mm^2
Tensile strength (numerical)	4.0	6.0	6.0	6.0	4.0	6.0	6.0	6.0	40.0
	Y^c_{11}	Y^c_{12}	Y^c_{13}	Y^c_{21}	Y^c_{22}	Y^c_{23}	Y^c_{31}	Y^c_{32}	Y^c_{33}
	N/mm^2	N/mm^2	N/mm^2	N/mm^2	N/mm^2	N/mm^2	N/mm^2	N/mm^2	N/mm^2
Compressive strength [6]	4.0	6.0	6.0	6.0	4.0	6.0	6.0	6.0	40.0
	ε^c_{11}	ε^c_{12}	ε^c_{13}	ε^c_{21}	ε^c_{22}	ε^c_{23}	ε^c_{31}	ε^c_{32}	ε^c_{33}
	adim.	*adim.*	*adim.*	*adim.*	*adim.*	*adim.*	*adim.*	*adim.*	*adim.*
Compressive maximum strain (numerical)	0.005	0.005	0.005	0.005	0.005	0.005	0.005	0.005	0.005

Table 3. Strength properties assumed for spruce wood in the numerical tests, which are named as tensile values Y^t_{ij} or compressive values Y^c_{ij}, i, j = 1, 2, 3 are the indices that correspond to longitudinal, radial and tangential fibre directions, respectively.

$\dot{\varepsilon}_{c0}$	a_1	a_2
adim.	*adim.*	*adim.*
1.0627E-7	4.7290	0.1458

Table 4. Creep parameters assumed for spruce wood in the numerical tests.

Another simulation has been reported as an example. It consists of the reproduction of the non-linear behaviour of a wood specimen under tension loading. The wood specimen has been dimensioned with a size of 20 mm × 10 mm × 240 mm, assuming that the orientation of the longitudinal fibre of wood is set along the loading axes. The simulation has been executed by using a finite element mesh of eight node hexahedral elements with 700 elements and 1188 nodes. In the test, the specimen has been restrained at its bottom surface edge and it has been subjected to an imposed displacement course at its top surface edge. The loading simulation has been carried out in control of displacement.

3. Results and discussion

The stress-strain data of both the experimental and numerical results for the specimen under compression loading have been plotted in **Figure 1**. One can view that the mechanical behaviour is properly predicted both for the elastic tract and for the yielding zone, although the last part of the elastic path, numerically predicted, stays slightly behind that measured.

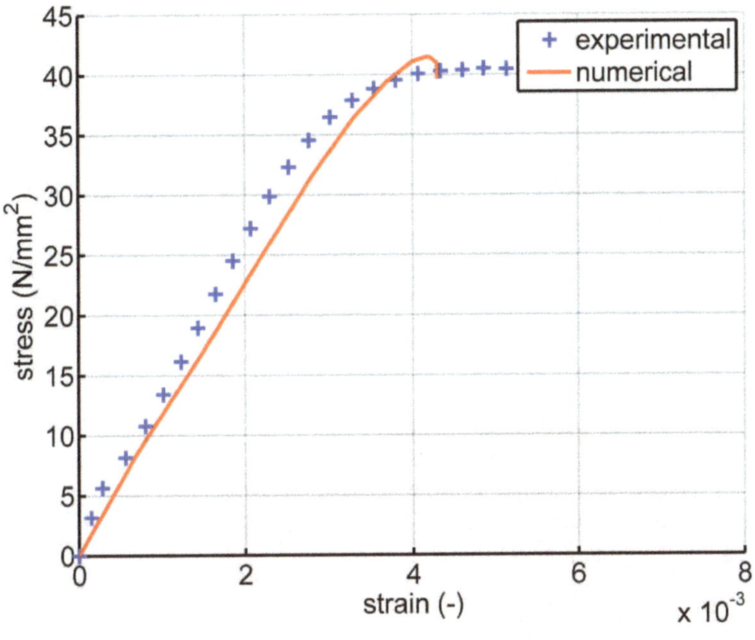

Figure 1. Comparison between numerical and experimental data of the stress-strain path for the compression test, under loading along the longitudinal axes.

Also the numerical results concerning the compression test are reported as a plot of coloured iso-map. These iso-maps show the intensity distribution of displacement, stress, strain and equivalent plastic (inelastic) strain along the longitudinal fibre, at some prefixed time steps (**Figures 2–5**).

In the plotted iso-map of the displacement field, along the longitudinal axes of the specimen (**Figure 2**), it is quite clear to observe the different levels of displacement along the specimen height, it varies from the maximum level at the top edge of the specimen to the minimum at the bottom edge, close to the restrains.

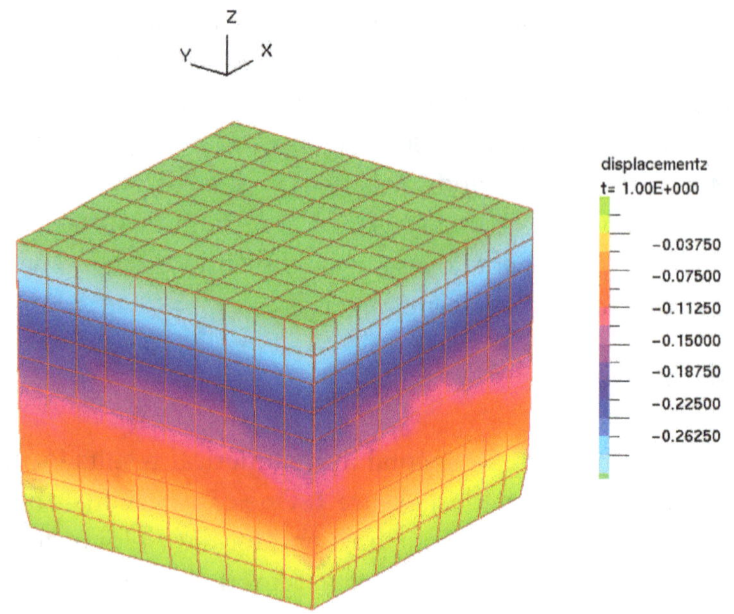

Figure 2. Compression test with the specimen loaded along the longitudinal axes. Plot of the iso-map of the displacement field.

Figure 3 shows the distribution of normal stress along the longitudinal axes of the specimen. The distribution is reasonably uniform with the exception around the edge restrain, where the marked colour variation points out the expected stress concentrations.

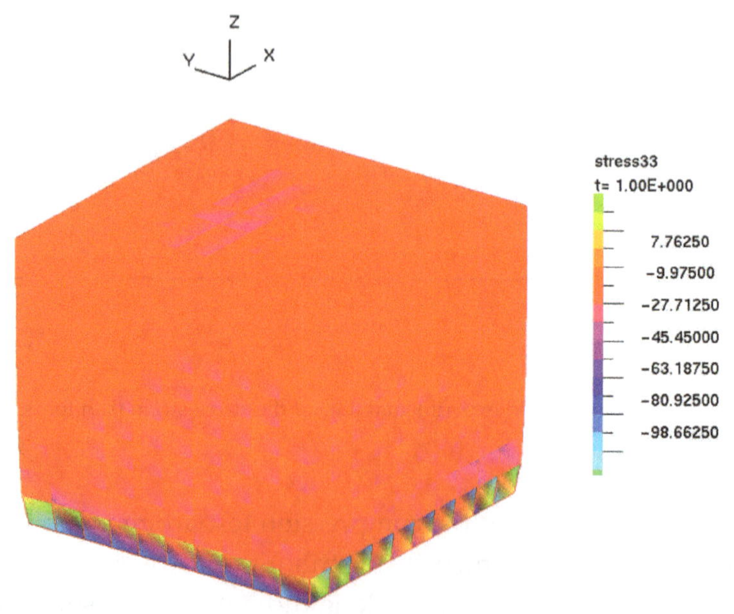

Figure 3. Compression test with the specimen loaded along the longitudinal axes. Plot of the iso-map of the stress field along the longitudinal axes.

Figure 4 displays the colour iso-map of the strain field along the longitudinal axes of the specimen. In this figure, the different levels of strain along the height of the specimen are quite evident, from a minimum at the bottom edge, which is cinematically restrained, to a maximum at the top, where the deformation increases from the inner side towards the outer side of the specimen.

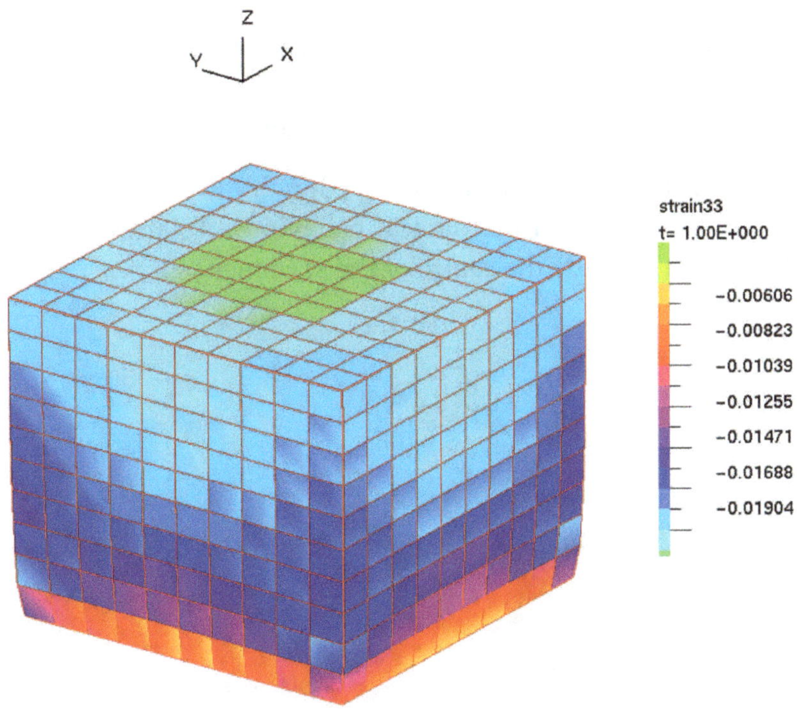

Figure 4. Compression test with the specimen loaded along the longitudinal axes. Plot of the iso-map of the strain field along the longitudinal axes.

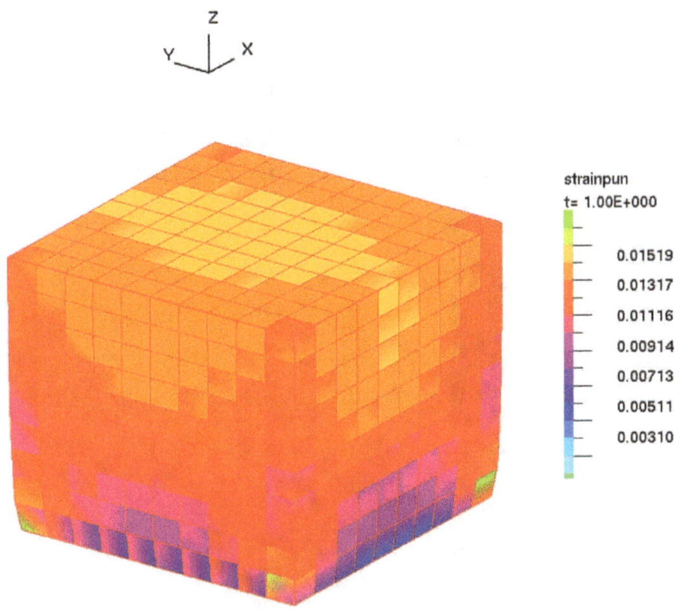

Figure 5. Compression test with the specimen loaded along the longitudinal axes. Plot of the iso-map of the equivalent plastic (inelastic) strain field.

Figure 5 shows the equivalent plastic strain distribution on the specimen shape, which is however comprehensive of other inelastic strains, such as creep. At the loading step illustrated in this figure, the plastic strain is quite extended on the whole of the specimen shape.

The numerical results for the specimen under tension loading are also reported as a plot of coloured iso-map. The iso-maps show the intensity distribution of displacement, stress and strain along the longitudinal fibre, at some prefixed loading stepping (**Figures 6–8**).

Figure 6 shows the displacement field along the longitudinal axes of the specimen. The distribution is quite coherent with a mechanical behaviour, it varies along the height of the specimen with a high level towards the centre of the specimen.

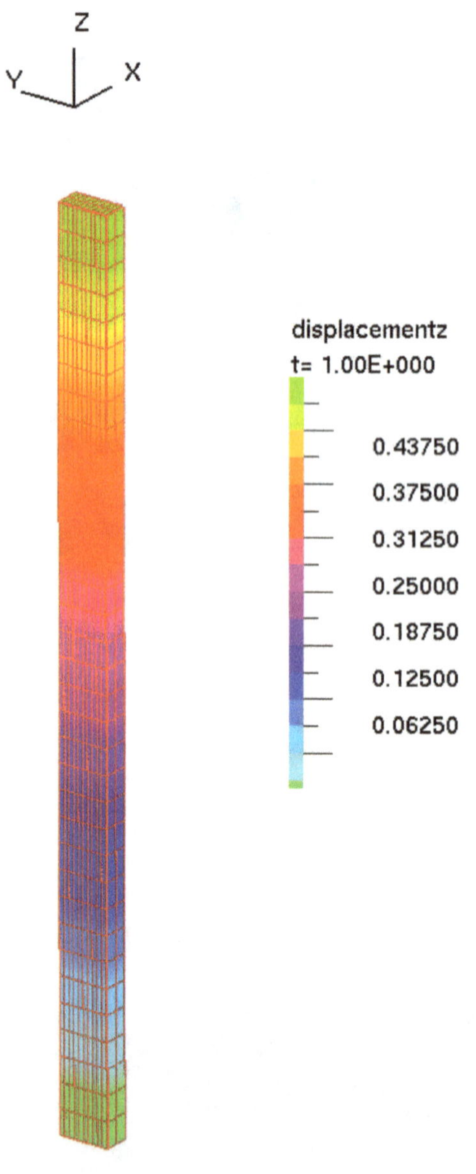

Figure 6. Tension test with the specimen loaded along the longitudinal axes. Plot of the iso-map of the displacement field.

In **Figure 7**, the stress distribution along the longitudinal axes of the specimen is displayed. The distribution is correctly rather uniform. It also presents stress concentrations near the bottom edge where the specimen is held back.

Figure 8 displays the strain iso-map along the longitudinal axes of the specimen. This distribution is quite uniform except near the bottom edge of the specimen.

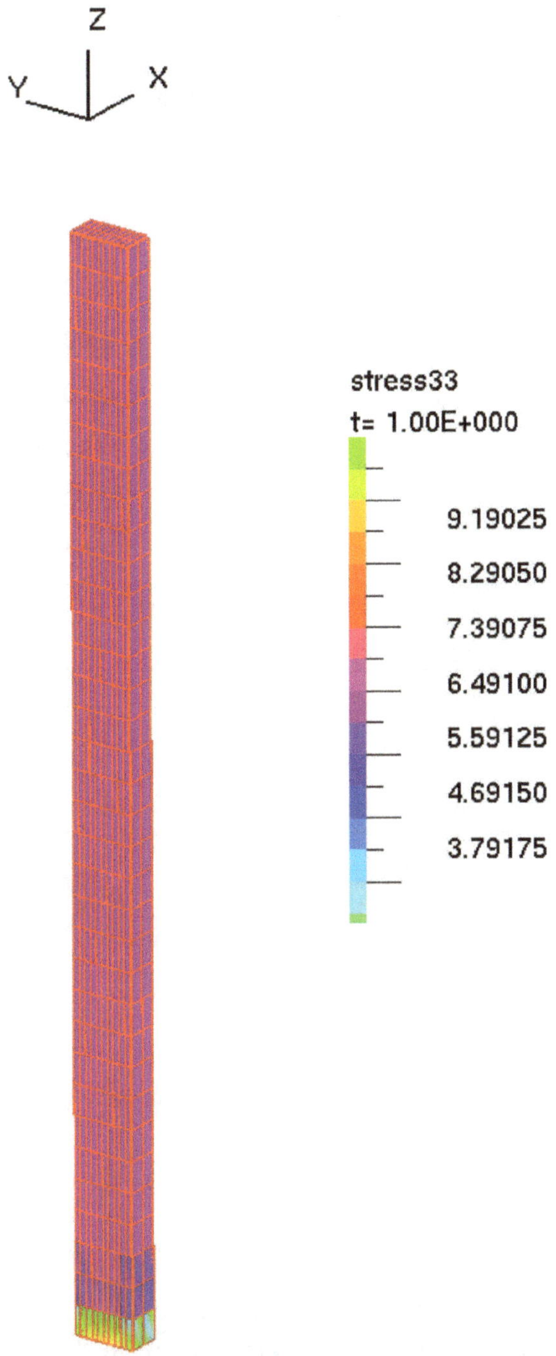

Figure 7. Tension test with the specimen loaded along the longitudinal axes. Plot of the iso-map of the stress field along the longitudinal axes.

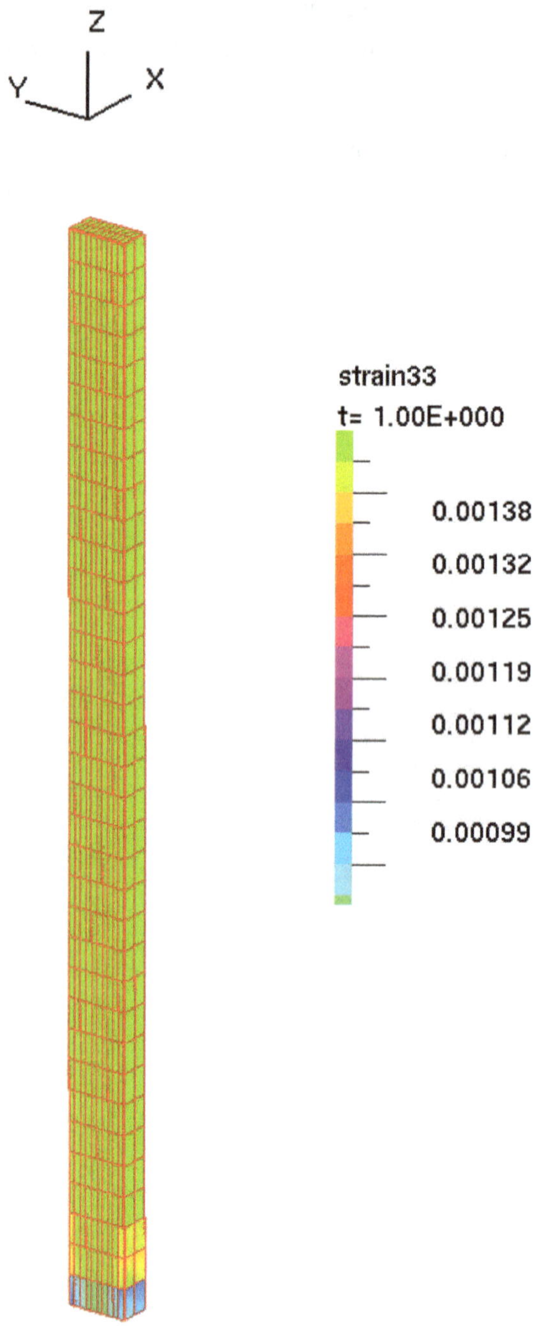

Figure 8. Tension test with the specimen loaded along the longitudinal axes. Plot of the iso-map of the strain field along the longitudinal axes.

4. Conclusions

Most common models, used to analyse the mechanical behaviour of timber or wood composite structures, mainly adopt a single equivalent yield surface to describe the non-linearity of the material. In the present approach a multi-surface yielding description of the non-linear behaviour of wood is proposed. It performs a general orthotropic elastic-viscous-plastic constitutive formulation, according to the finite strain theory. Within a continuum mechanics approach, the proposed three-dimensional FEM model is capable to represent elastic-viscous-plastic in compression and elastic-brittle in tension behaviour of wood.

On the basis of a preliminary comparison of the numerical results with experimental data, the model has demonstrated to be valuable to represent the main mechanical behaviour of a test specimen. The examples presented here illustrate that the FEM model can be adequate to examine the mechanical behaviour of wood under displacement loading both in linear and non-linear situations. The model has been built within a general purpose FEM code, its formulation being quite general it is almost powerful to describe complex states in wood materials such as anisotropy, elasticity and visco-plasticity, brittle crack occurring in the timber structure. However, the validation of the model with the comparison between numerical and experimental results can be carried out in a future work

Acknowledgements

This work has been supported by the University of Basilicata under the research Project 'RIL 2015'. This support is gratefully acknowledged.

Author details

Vincenzo De Luca

Address all correspondence to: vincenzo.deluca@unibas.it

Dipartimento DIS, Università degli Studi della Basilicata, Potenza, Italy

References

[1] De Luca V., Marano C. Prestressed glulam timbers reinforced with steel bars. Construction and Buildings Materials. 2012; **30**:206–217. DOI: 10.1016/j.conbuildmat.2011.11.016.

[2] Patton-Mallory M., Cramer S. M., Smith F. W., Pellicane P. J. Nonlinear material models for analysis of bolted wood connections. Journal of Structural Engineering. 1997; **123(8):**1063–1070.

[3] Perkins R. W. Concerning the mechanics of wood deformation. Forest Production Journal. 1967; **17**(3):55–67.

[4] Bodig J., Jayne B. A. Mechanics of wood and wood composites. New York, USA: Van Nostrand Reinhold Company; 1982.

[5] Dinwoodie J. M. Timber—its nature and behaviour. New York, NY: Van Nostrand Reinhold Company; 1981.

[6] De Luca V., Sabia D. Mechanical compression tests to model timber structures behaviour. In: A. A. Mammoli and C. A. Brebbia, editor. Computational Methods and Experiments

in Materials Characterization III. WIT Transaction on Engineering Sciences; 13 June 2007 through 15 June 2007; Bologna (Italy). Southampton: WIT Press. Southampton; 2007. Volume 57 pp. 273–278. DOI: 10.2495/MC070271.

[7] Hankinson R. L. Investigation of Crushing Strength of Spruce at Varying Angles of Grain. Air Service Information Circular No. 259, U. S. Air Service; 1921.

[8] Goodman J. R., Bodig J. Orthotropic strength of wood in compression. Wood Science. 1971; **4**(2):83–94.

[9] Tsai S. W., Wu E. M. A general theory of strength for anisotropic materials. Journal of Composite Materials. 1971; **5**:58–80.

[10] Tabiei A., Wu J. Three-dimensional nonlinear orthotropic finite element material model for wood. Composite Structures. 2000; **50**:143–149.

[11] Schmidt J., Kaliske M. Models for numerical failure analysis of wooden structures. Engineering Structures. 2009; **31**:571–579. DOI: 10.1016/j.compstruc.2011.06.004.

[12] Oudjene M., Khelifa M. Finite element modelling of wooden structures at large deformations and brittle failure prediction. Materials and Design. 2009; **30**:4081–4087.

[13] Dahl K. B., Malo K. A. Nonlinear shear properties of spruce softwood: numerical analyses of experimental results. Composites Science and Technology. 2009; **69**:2144–2151.

[14] Oudjene M., Khelifa M. Elasto-plastic constitutive law for wood behaviour under compressive loadings. Construction and Building Materials. 2009; **23**:3359–3366.

[15] Khennane A., Khelifa M., Bleron L., Viguier J. Numerical modelling of ductile damage evolution in tensile and bending tests of timber structures. Mechanics of Materials. 2014; **68**:228–236.

[16] De Luca V., Della Chiesa A. A creep non-linear FEM analysis of glulam timber. Mechanics of Advanced Materials and Structures. 2013; **20**:489–496. DOI: 10.1080/15376494.2011.627643.

[17] Moayyed M. Y, Taheri F. Creep response of glulam reinforced by a novel pre-stressed FRP-wood composite system. 11th World Conference on Timber Engineering 2010 (WCTE 2010), Trentino, Italy, 20–24 June, 2010.

[18] Bathe K. J. Finite element procedures. Upper Saddle River: Prentice Hall; 1995.

[19] Zienkiewicz O., Taylor R. The finite element method. Volume 1: The basis. Oxford: Butterworth-Heinemann; 2000.

[20] Zienkiewicz O., Taylor R. The finite element method. Volume 2: Solid mechanics. Oxford: Butterworth-Heinemann; 2000.

[21] Carosio A., Willam K., Etse G. On the consistency of viscoplastic formulations. International Journal of Solids and Structures. 2000; **37**:7349–7369.

[22] De Angelis F. On the structural response of elasto/viscoplastic materials subject to time-dependent loadings. SDHM Structural Durability and Health Monitoring. 2012; **8** (4):341–358.

[23] De Angelis F. Numerical algorithms for j2 viscoplastic models. Advanced Materials Research. 2012; **567**:267–274. DOI: 10.4028/www.scientific.net/AMR.567.267

[24] De Angelis F. Computational aspects in the elasto/viscoplastic material behavior of solids. Advanced Materials Research. 2012; **567**:192–199. DOI: 10.4028/www.scientific.net/AMR.567.192

[25] De Angelis F. On constitutive relations in non-smooth elasto/viscoplasticity. Advanced Materials Research. 2012; **566**:691–698. DOI: 10.4028/www.scientific.net/AMR.566.691

[26] De Angelis F. Computational issues in rate-dependent plasticity models. Advanced Materials Research. 2012; **566**:70–77. DOI: 10.4028/www.scientific.net/AMR.566.70.

[27] Simo J. C., Ortiz M. A unified approach to finite deformation elastoplastic analysis based on the use of hyperelastic constitutive equations. Computer Methods in Applied Mechanics and Engineering. 1985; **49**:221–245.

[28] Moran B., Ortiz M., Shih. C. F. Formulation of implicit finite element methods for multiplicative finite deformation plasticity. International Journal for Numerical Methods in Engineering. 1990; **29**:483–514.

[29] Wang W. M., Pozivilova A., Sluys L. J. Implicit algorithms for finite deformation viscoplasticity. European Congress on Computational Methods in Applied Sciences and Engineering, ECCOMAS 2000, Barcelona 11–14 September, 2000.

[30] Wang W. M., Sluys L. J. Formulation of an implicit algorithm for finite deformation viscoplasticity. International Journal of Solids and Structures. 2000; **37**: 7329–7348.

[31] García Garino C., Ribero Vairo M. S., Andía Fagés S., Mirasso A. E., Ponthot J.-P. Numerical simulation of finite strain viscoplastic problems. Journal of Computational and Applied Mathematics. 2013; **246**:174–184. DOI: 10.1016/j.cam.2012.10.008.

[32] Cuitino A. M., Ortiz M. A. A material-independent method for extending stress update algorithms from small-strain plasticity to finite plasticity with multiplicative kinematics. Engineering Computations. 1992; **9**:437–451.

[33] Rodríguez-Ferran A., Pegon P., Huerta A. Two stress update algorithms for large strains: accuracy analysis and numerical implementation. International Journal for Numerical Methods in Engineering. 1997; **40**(23):4363—4404.

[34] Lubliner J Plasticity theory. First ed. New York, USA: Macmillan Publishing Company; 1990.

[35] Eidel B., Gruttmann F. Elastoplastic orthotropy at finite strains: multiplicative formulation and numerical implementation. Computational Materials Science. 2003; **28**:732–742.

[36] Holmes D. W., Loughran J. G. Numerical aspects associated with the implementation of a finite strain, elasto-viscoelastic–viscoplastic constitutive theory in principal stretches.

International Journal for Numerical Methods in Engineering. 2010; **83**:366–402. DOI: 10.1002/nme.2850.

[37] Schröder J., Gruttmann F., Löblein J. A simple orthotropic finite elasto–plasticity model based on generalized stress–strain measures. Computational Mechanics. 2002; **30**:48–64. 10.1007/s00466-002-0366-3.

[38] Forest Product Laboratory (FPL). Wood handbook-wood as an engineering material. General Technical Report FPL-GTR-113. Madison (WI, USA): Department of Agriculture; 1999.

[39] Lekhnitskii S. G. In: Brandstatter J. J., editor. Theory of elasticity of an anisotropic elastic body. San Francisco, CA: Holden Day, Inc.; 1963.

10

Exterior Wood Coatings

Mojgan Nejad and Paul Cooper

Abstract

In addition to aesthetic appeal, coatings are designed to protect the wood from weathering degradation in outdoor conditions. This chapter gives an overview of the effects of the main coating components, coating properties, wood properties and treatments on coated wood performance in service. Understanding how different type of resins, pigments, solvents, and other major additives affect coating performance on wood, helps coating formulators to develop more durable coatings. It is beneficial for both wood scientists and coating chemists to learn which properties of coatings have the highest impact on predicting their service lives when exposed to weathering. For instance, measuring glass transition temperature (Tg) defines the degree of flexibility of a coating. Since wood swells and shrinks due to moisture uptake and subsequent drying, flexibility of a coating plays a critical role in defining its durability on wood in exterior conditions. Similarly, learning how preservative treatment or a new modification technique changes the surface properties of wood will help coating formulators to adjust properties of coatings in way that would have better adhesion and performance on that specific modified wood. Also, the effects of these factors in increasing service life of exterior wood coatings are discussed.

Keywords: exterior, wood coating, weathering, treatment, performance and durability

1. Introduction

The performance of coatings in exterior applications depends on many factors such as wood species and their physical properties, coating types and properties, application procedures, and exposure conditions. In North America, cedar (*Thuja* species), spruce (*Picea* sp.), pine (*Pinus* sp.) and fir (Abies sp.) are the most commonly used wood species in exterior building constructions. Among these, cedar has very low density, and it is known as a naturally durable wood species. The fungicidal effect comes from extractives that exist in heartwood of cedar, which provide protection against decay fungi and insect. Cedar boards are available for decking, fencing and siding, but they cost more than other softwoods (spruce, pine and fir that are generally called SPF). SPF lumber is usually treated with either preservative chemicals or modified by high temperature if intended for outdoor or contact ground applications. More detail on effects of different treatments on changing the surface properties of wood is discussed later in the chapter.

In addition to deterioration caused by decay fungi and insects, wood-exposed outdoors go through some physical and chemical changes known as weathering. Weathering degradation mainly affects the surface of the wood, except when the checks that form on the wood surface expand deeper into the boards. Weathering is caused by exposure to UV and visible radiation, moisture (rain, snow and dew) and wind. UV and visible radiation degrade wood components, especially lignin, causing the surface to turn grey and fibres to loosen and be abraded or removed by insects such as wasps for their nests. Precipitation contributes to leaching of wood extractives and their migration to the wood surface where they initially darken the wood but eventually contribute to greying of the surface as they leach out. Changes in moisture content cause swelling and shrinking leading to stresses and checking of the surface. Repeated freezing and thawing of liquid water in the wood cells result in additional stresses in the wood. Wind and wind-driven particles scour the wood surface contributing to uneven wear in the earlywood and latewood. Some also consider the staining and surface deterioration by mould, soft rot and decay fungi on the wood surface as a biological component of weathering.

Application of surface coatings on wood is one of the best ways to reduce weathering degradation. However, depending on the severity of exposure and coatings' type and properties, refinishing is needed every few years. For instance, solid colour stains and paints have much longer service life than transparent or clear formulations. This is due to the fact that clear or transparent formulations do not have pigments to absorb the UV rays; thus, the UV can pass through the clear coating layer and cause oxidation of lignin. Lignin degradation in one of main reasons for wood colour change and contributes to increased water uptake and defibrillation of cellulose fibres from the wood surface. Thus, if we can find a good way to prevent lignin degradation, for instance by having high pigment concentration or other UV absorbers, then the coating will last longer. The main drawback of high pigment is that it will hide the beauty of the wood grain.

Formulating coatings for wood, as a biological material with huge variability, is not an easy task, especially when the coatings film has to have excellent weather resistance and provide exceptional protection for wood against weathering factors. Exterior wood coatings have very complex formulations and are categorized in different ways, for example, based on the carrying media, that is, water based versus solvent based, on the amount of pigment that is added to the formulation, that is, solid, semi-transp`arent and transparent coatings, and based on whether they form a film (film forming) or are penetrating stains. Most coating formulations are made of the following main components: (1) resin or binder, (2) solvent which could be organic solvent or water, (3) pigments (transparent coatings either do not have any pigment or only contain nanopigments) and (4) other additives. Additives as the name implies are chemicals that are added in very small quantities but play significant roles in defining properties and performance of the coatings on wood products.

Effects of wood properties, wood treatment, weathering and coating properties, and how all of these can affect coating performance are discussed in more detail in the chapter. We hope that by the end of this chapter, readers will have a better understanding of factors affecting coating performance in exterior applications and how we can improve the durability and service life of coatings.

2. Weathering of wood

Weathering is mainly a surface phenomenon and the grey layer of weathered wood is only about 125-μm thick [1]; however, checking and cracking of the wood may occur if wood is left uncoated for an extended period of time. In general, weathering is a combination of degradation by solar radiation (ultra violet, visible light and infrared), moisture changes (rain, dew, humidity and snow), oxidation and temperature effects (heat and freezing). Among these factors, UV degradation and moisture effects are the most influential factors on performance of coated wood.

The majority of light (80–95%) is absorbed by lignin in wood [2]. Lignin absorbs light between 200 and 400 nm with a strong absorption peak at 280 nm [1]. Photo-oxidation of lignin will cause the formation of phenolic radicals, which turn quickly to ortho and para quinonoids chromophoric groups [1]. Formation of these yellow to brownish compounds causes the initial dark brown colour of wood during weathering. Further exposure of wood surfaces to rain leaches out degraded lignin from the surface and leaves cellulose fibre exposed, a phenomenon known as "roughening". The loss of a hydrophobic compound (lignin) from the wood surface not only loosens the fibres but also reduces the water repellency of the wood.

In addition to the effects of moisture on swelling and shrinking of the wood discussed above, moisture also helps the radicals formed during UV exposure to diffuse deeper into the wood. Although the UV light can only penetrate up to 75 μm in the wood [3], the thickness of the brown layer beneath the grey wood has been reported to be up to 2500 μm [1]. Visible light has higher penetration (200 μm) than UV but does not have sufficient energy (70 kcal/mol) to initiate radical formation or cleave chemical bonds [1]. Thus, this brown discoloration of wood is not caused by visible light confirming that the radicals formed at the surface are transferred inside the wood by water.

In addition to discoloration and defibrillation, exterior woods develop extensive cracking, checking and mould or mildew growth on the surface. The mildew growth that usually appears as black spots on the wood or coating surfaces are different from decay fungi in that they consume nonstructural components such as sugars and starches and do not degrade the major components of wood (cellulose, hemicelluloses and lignin). The mildew can be removed with a dilute solution of sodium hypochlorite. Most wood coating formulations contain mildewcides. Checking of the wood is much more difficult to prevent. When wood adsorbs moisture, it swells, and as it dries, it shrinks; when the stresses exceed the elasticity and strength of the wood [4], small cracks will develop which later expand to checks if wood remains exposed to outdoor conditions. Evans et al. [5] reported that samples that were exposed to UV-rays had higher surface checking than those that were protected by filter. The study showed that direct exposure to UV increases the surface checking as a result of lignin degradation.

3. Factors affecting performance of exterior coatings on wood

3.1. Effects of wood properties

Variability in wood contributes to its aesthetic appeal when compared with other building materials like cement, steel or plastic. However, this variability creates many challenges for coating chemists. Wood properties not only differ significantly among different wood species and especially between hardwoods and softwoods, they also vary within the same board [6]. Most hardwoods have relatively higher density than softwoods and have coarser grain because of the relatively large vessels or pores. Within the same board, there is a significant variation in the density and permeability of the wood depending on the latewood to earlywood ratio and the relative amounts of sapwood, outer heartwood, and pith-associated heartwood or juvenile wood. For example, latewood of southern pine is almost three times as dense as the earlywood [7]; thus, the wider the latewood bands, the higher the density of the section of the board (**Figure 1**). In general, coatings perform better on low-density wood such as cedar and redwood than on higher density woods like Douglas fir and southern pine [8] because low-density woods have lower shrinking and swelling than high-density wood [9]. Swelling happens as a result of water adsorption in the cell walls of wood, which continues only until wood reaches its fibre saturation point at about 30% moisture content (dry mass basis) [10]. The thicker cell walls of latewood hold more moisture in a given volume and so swell or shrink more than thinner earlywood cell walls. This creates stresses in a coating which explains why coatings start to fail in the latewood and then failure progresses into earlywood [11, 12]. In flat-sawn lumber where the tangential surface is exposed on the face of the board, wood has wider latewood and earlywood bands exposed than on quarter-sawn lumber. Thus, coatings show higher durability performance when applied on the radial faces of quarter-sawn boards [9].

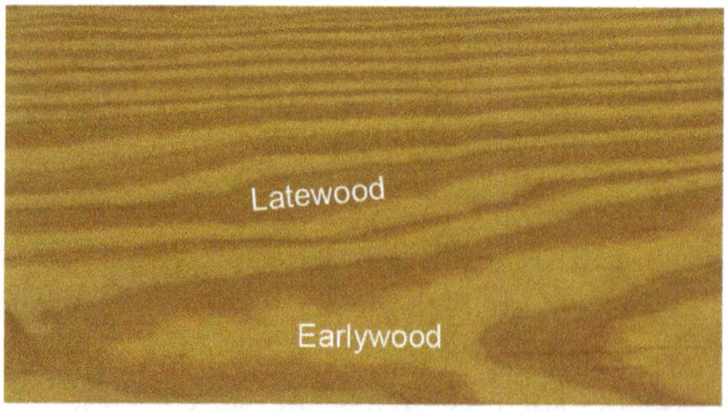

Figure 1. Image of flat-grained southern yellow pine wood showing distinctive wide earlywood and latewood sections.

Moisture content of wood is another important property that affects coating performance on wood. Wood swells when it absorbs moisture and shrinks when it dries; these dimensional changes are more pronounced in latewood than in earlywood. Shrinking and swelling of the wood constantly cause stresses in the coating film. The moisture content of green wood can be more than 150% (dry wood basis), but above the fibre saturation point (FSP ≈ 30%), wood does not absorb any more water in its cell walls and is in its fully swollen or green condition and dimensionally stable [10]. Applying coatings on wet wood may cause early failure of the coating because of initial adhesion issues, shrinking of wood underneath, and migration of moisture out of the surface as wood dries. This is especially problematic for film-forming low-permeability paints. If a coating is applied to wood under these moisture conditions, it must have good water vapour permeability [13] to allow the absorbed water to escape without causing coatings to blister. If the water stays in wood, it will create suitable conditions for growth of decay fungi inside the wood. In service, wood adsorbs or desorbs moisture in response to ambient relative humidity conditions, and eventually, it reaches its equilibrium moisture content (usually around 12–15% MC). It is recommended to apply a coating when wood reaches moisture content close to its average equilibrium moisture content in service [14].

In some wood species, natural extractives negatively affect coating appearance and performance. Extractive bleeding is caused by migration of water-soluble extractives or thermally induced movement of water-insoluble resin or pitch to the surface of the coating. Resin (from wood) usually leaves a yellow stain which is sticky to touch. Pitch (a mixture of rosin and turpentine) problems are mostly associated with pine, spruce and fir [15]. Proper kiln drying of the wood to evaporate the turpentine and set the resin, application of a sealant such as shellac to knots or resin pockets and cleaning of the wood surface with turpentine before application of the coating have been reported to resolve or minimize these issues [12].

For exterior applications, coatings have better adhesion to rough sawn woods [15] than to planed or sanded wood. This is probably due to the fact that rough surfaces have more anchoring sites for mechanical interlocking, thus improving adhesion of the coating to the wood. It is important to remember that there should not be any saw dust remaining on the wood surface either after sanding or planing, because then the coating will adhere to the saw dust not the wood. The surface of the wood after sanding can be easily cleaned using a damp cloth to remove extra saw dust from the surface and leave the pores open for possible coating penetration.

3.2. Effects of wood treatments

Unlike weathering, decay fungi and insects can completely deteriorate wood and cause serious structural damage. To prevent or delay deterioration of the wood, most wood products should be protected from decay fungi and insects if intended for exterior applications [16]. These protections can be divided into two categories: chemical and thermal treatments. In chemical treatments, either preservative chemicals (classed as pesticides) are impregnated into the wood, or different chemicals are used to modify the wood to make it less susceptible to biodegradation. Preservative formulations usually include fungicide(s) and insecticide(s), which are added to the wood by using high pressure and vacuum applications [16]. The degree of treatment depends on the formulation, preservative loading and treatability of different wood species. In a three-year natural weathering study, we observed that overall preservative treatment-enhanced coating performance [17]. All the tested coatings had better general performance on all preservative treated wood than they did on untreated southern pine samples. After 3 years of natural exposure, coatings had less surface erosion and peeling on preservative-treated wood than they had on untreated wood as shown in **Figure 2**.

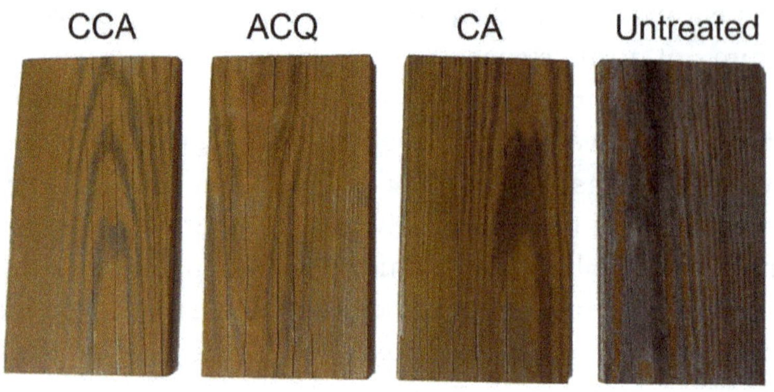

Figure 2. Image of treated and untreated southern pine samples coated with an alkyd-acrylic water-based coating after 3 years of natural weathering in Toronto, Canada.

Up until 2005, many decks and fences were treated with waterborne-chromated copper arsenate (CCA) formulation, but due to concern about leaching of arsenic and chromium from CCA-treated wood, the preservative industry voluntarily shifted to other copper-based or organic preservatives. In North America, waterborne alkaline copper quaternary ammonium compound (ACQ), copper azole (CA) and micronized copper are dominant formulations in the residential market now. Although oil-based preservative formulations such as pentachol-orophenol (PCP) and creosote are used in several industrial products, coatings are mainly applied to above ground-treated wood products such as fences, decks and siding that are treated with water-based formulations.

CCA was the main formulation for treating woods for about 50 years almost from 1953 when CCA was first standardized by AWPA, until the voluntary phase out from residential application in 2005 [18]. During this time, many exterior coatings were formulated and tested on CCA. Coatings have been reported to generally perform better on CCA-treated woods than untreated woods [17, 19]. We observed that CCA-treated wood had much better water repellence performance compared to ACQ and CA [17]. **Figure 3** shows the moisture content changes of the different coated treated woods during 3 years of natural weathering. As can be seen, the alkaline Cu-based preservative-treated wood (ACQ and CA) had higher water uptake than even untreated wood [17]. This is a good example of why coating chemists should understand how different treatments change the properties of the wood. Many coatings that were formulated for exterior applications before 2005 were tested on CCA-treated wood [17]. They did not show the same level of performance on the replacement alkaline Cu-based preservative-treated wood during 3 years of exposure to natural weathering in Toronto, Canada [17]. Thus, with any new formulation or changes in wood treatments, coating industries should consider modifying their formulations based on how the treatment has changed the surface properties of the wood. In this case, in order for coatings to have high durability, they should have higher water repellence performance than those formulated previously for residential decking and fences that were designed for CCA. Both CCA and alkaline Cu-based preservatives provide some degree of UV protection by modifying lignin [20] and reducing the colour change of coated-treated wood in comparison with untreated wood [17], but they have completely different performance when it came to water repellency.

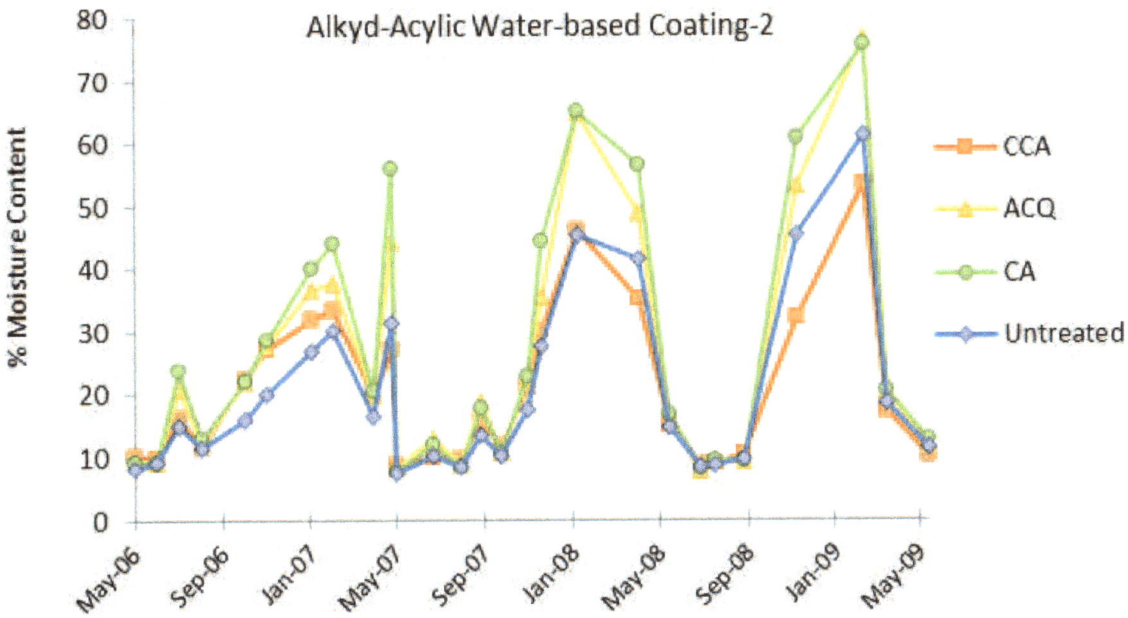

Figure 3. The average percent moisture content changes of coated-treated southern pine samples during 3 years of natural weathering in Toronto, Canada.

Another issue associated with preservative-treated wood is possible leaching of preservative components from wood in service [21–23]. We also studied effect of coatings in reducing leaching of heavy metals (As, Cr and Cu) and found that application of one coat of semitransparent stain could effectively reduce the leaching of these preservative components on average by 60% [24]. One of the most interesting points was that Cu leaches at a really high rate on uncoated wood during the first 6 months of exposure. We observed that even when the coatings start failing and eroding from the wood surface, the rate of Cu leaching from coated wood never reached to the same level as of uncoated samples (**Figure 4**). Further in-depth analysis of samples showed that application of coating during early stage of exterior exposure help fixation of copper in wood, by protecting it from the initial high rate of leaching until it could react more completely with the wood.

Multivariate analysis of data obtained during 3 years of natural weathering of performance of coatings on different kinds of treated wood showed that preservative treatments (CCA, ACQ and CA) had a greater effect than any other coating properties such as resin type, base (water-based or solvent-based), flexibility or film thickness on coating performance on wood [25].

Another modification method used at the commercial scale is protecting wood by changing the surface chemistry of wood by acetylation. Acetylation reduces the hydrophilicity of the wood, thus its equilibrium moisture uptake to a level that would not be favourable for decay fungi [26]. In this approach, acetic anhydride treatment will replace hydrogen in hydroxyl functional groups of mainly hemicelluloses in wood with acetyl groups (CH_3CO–) [27]. Acetylation of wood is associated with weight gain of treated wood of about 30–40% [26]. Acetylated wood has been reported to protect the wood when the percent weight gain is more 20% after treatment [28]. This higher dimensional stability of acetylated wood automatically reduces stresses applied to the coating's film, resulting in overall better coating performance. However, coated-acetylated wood was reported to have significant surface mildew growth on the finished surface after 3 years of natural weathering which probably impaired the effectiveness of the coatings [29]. Acetylation of wood was shown to improve colour stability of wood by preventing lignin degradation when modified woods were exposed to UV radiation [27, 30, 31].

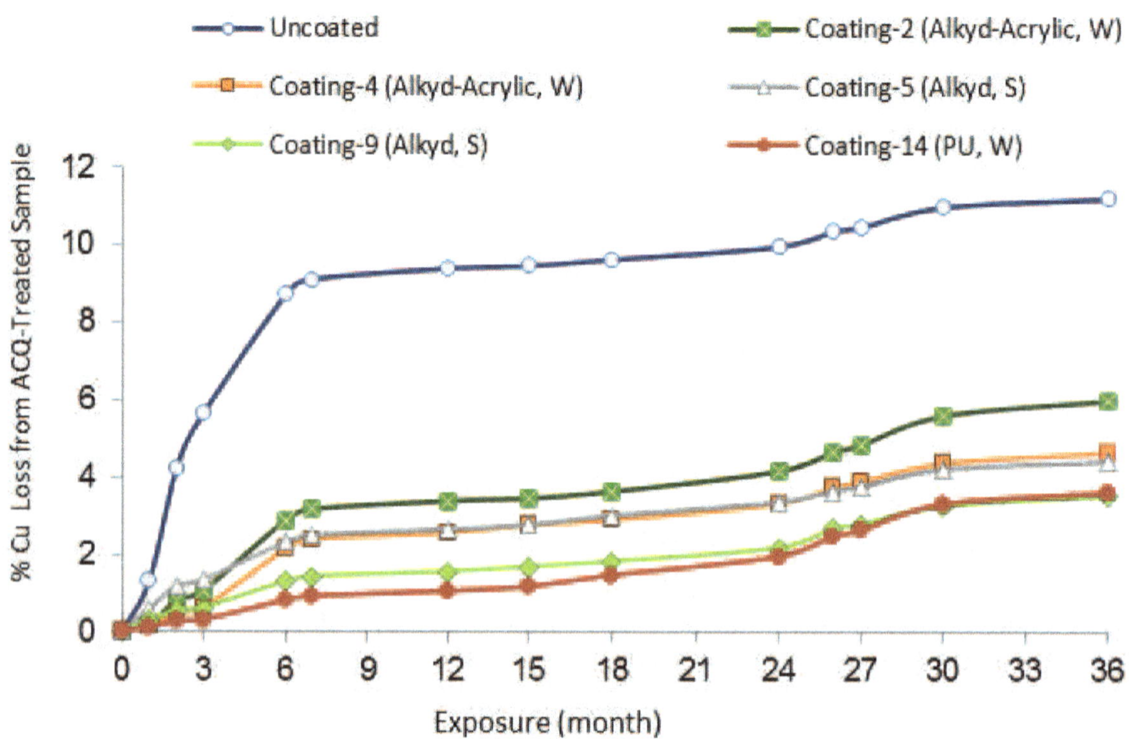

Figure 4. Effect of semi-transparent stain coatings on copper leaching reduction from ACQ-treated wood samples during 3 years of natural weathering exposure in Toronto, Canada (W = water based, S = solvent based).

Thermal modification is another treatment method, which has growing market acceptance, especially in Europe. In thermal treatment, wood is heated at about 200°C under oxygen-free environment for a few hours (3–7 h) using either steam or hot oil as media for continuous heat transfer [32]. Thermally modified wood, like acetylated wood, has improved dimensional stability due to lower water uptake as a result of decrease in available hydroxyl functional groups in hemicelluloses. Protection provided by thermal treatment is not as effective as preservative treatment; therefore, the thermally modified wood is not recommended for ground contact applications [33]. Since during thermal modification wood is only subjected to high temperature and no chemicals are used, thermally modified wood is considered as a more environmentally friendly product than preservative-treated wood. It also has the advantage of complete treatment of the wood, even in species that cannot be penetrated effectively by preservatives or acetylation chemicals. The most important benefit of the thermal modification process is the improved dimensional stability of the wood [26, 33, 34]. The chemical changes that occur during heat treatment such as reduced hydroxyl content and higher degree of cellulose crystallinity [35] increase the hydrophobicity of wood, thereby improving the water repellence performance of coated heat-treated wood [36].

3.3. Effects of coating components

Formulation of coatings is very complex, and we can only cover a few major components and their potential interaction effects with wood in this chapter. Coatings are generally made of four main components: (1) resins or binders, (2) pigments, (3) solvents and (4) other additives. Different coating formulations may not have all these four main components. For instance, powder coatings do not have any solvents, but they are mostly used for metals which can be electrically charged. Also, clear and transparent coatings usually do not have any pigments in their formulations. For exterior wood applications, we are mainly dealing with water-based and solvent-based coatings, UV-curable coatings are becoming more popular for interior wood flooring but not yet for exterior applications. Most coatings for exterior woods are applied at the site, except for some siding products, window frames and doors which are factory primed (mostly white) and occasionally finish coated.

3.3.1. Resins

In any coating formulation, the main component is the resin (or "binder"). Resin governs to a large extent the major properties of the coating [37]. Exterior wood coatings are largely made with alkyd, acrylic or polyurethane resins. Recently, combinations of two binders that are called hybrid or core-shell systems are also used in wood coating formulations like alkyd-acrylic, acrylic-urethane, and alkyd-urethane.

Alkyd resins, which are modified-natural polyester resins, are one of the first synthetic resins used in surface coatings and are the most commonly used binders in exterior wood coating formulations. Alkyd is derived from polyols (glycerol), fatty acid oil (linseed oil or vegetable oil), and dibasic acid (phthalic anhydride) [38]. Alkyd-based coatings have generally poorer outdoor durability than acrylics and polyurethanes but have fewer film defects and lower costs [37]. Alkyd coatings are cured through an oxidation process [39], and sometimes it takes months to cure in outdoor exposure. This slow curing will affect the initial water repellence performance of an alkyd-based coating, but it improves over time as the resin cures [17]. Alkyd resins have very low molecular weights and therefore have potential to penetrate into wood cell walls when they are formulated at low viscosity range that can penetrate the wood surface.

Acrylic resins are prepared by chain-growth polymerization of various (meth) acrylic monomers [37]. They have excellent exterior durability [37] and are especially known for their superior UV resistance and resistance to yellowing [40]. Acrylics have high molecular weight (usually 75,000 or higher) which would result in higher film strength [37, 41]. But resins with such a high-average molecular weight cannot penetrate into wood cell walls. The largest polymeric compound that is reported to penetrate into wood cell wall is polyethylene glycol of 20,000 [42].

Polyurethanes (PU) are made of a polyol (R-OH) and isocyanate (R-NCO) forming urethane or carbamate linkages (>NH–CO–O<) through a condensation polymerization reaction [37]. Polyurethanes have high chemical resistance and are well known for their superior abrasion resistance [37]. They are mostly used as topcoats for flooring, cabinets and automotive applications [37]. Polyurethane resins generally have low molecular weights which would enable them to penetrate into wood cell walls. Free isocyanate of PU resins can easily form chemical bonds with hydroxyl groups in the wood. In addition, coatings made with aliphatic diisocyanates with hindered amine have exceptional exterior durability [37]. There are both solvent-based and water-based formulations of all three resins (alkyd, acrylic and PU) in the market.

3.3.2. *Pigments*

Pigments are important components of the exterior coatings, providing colour, opacity and UV protection. Inorganic pigments are the most universal pigments used in the formulation of exterior wood coatings. Iron oxide (red, yellow, brown and black) is frequently used in wood stains, whereas titanium dioxide (TiO_2) is used for white or other light coloured paints. TiO_2 has excellent hiding power, but it is relatively expensive compared to other white pigments. In most pigmented formulations, extenders and fillers are added to formulations; these commonly include calcium carbonate ($CaCO_3$) or talc (magnesium silicate, $H_2Mg_3(SiO_3)$). $CaCO_3$ has much lower hiding power than TiO_2, but it is more affordable. In transparent coatings, nanopigments are also used to protect the wood from UV degradation and/or improve coating scratch resistance (by using Al_2O_3) while not blocking the beauty of the wood grain [43, 44].

3.3.3. *Solvents*

Except for powder coatings and high solid formulations, about 30–50% of most coatings is comprised of solvents. Solvents can be either organic liquids or water. The most commonly used organic solvents in wood coating formulations are mineral spirits, methyl ethyl ketone, butanol, propyl alcohol, ethyl acetate, acetone, toluene and xylene [4]. Solvents are added for various purposes such as dissolving the resin and adjusting viscosity to a range that helps the application of coating either by brush or spray. In every coating formulation, a mixture of different solvents is used to optimize the required properties such as improving spray application and reducing the drying time. To meet environmental regulations for reduced amounts of volatile organic compounds (VOC) in coatings, water-based formulations were developed. Water-based formulations are formulated either with very low amounts of organic solvents or none with zero VOC.

3.3.4. *Additives*

As their names implies, additives are substances that are added to the coating in very small quantities to impart specific properties. Additives can be divided into various groups such as wetting and dispersing agents, defoamers, flattening agents, rheology modifiers, light stabilizers, dries, accelerators, and biocides [38, 45]. In water-based formulations, low evaporation rate of water compared with organic solvents, especially aromatic solvents, is a major challenge. To reduce drying time and ensure a defect-free dry film, co-solvents, coalescing agents or driers (mostly for alkyd resin) are used in the formulations [40]. Getting into detail of coating formulation and additives is beyond the scope of this chapter. However, it should be noted that even small amounts of additives can significantly change properties of coatings. For instance, we observed that addition of 0.1% a surfactant and wetting agent significantly reduced the surface tension of a water-based PU resin from 44 to 29 mN/m, minimized initial foam formation and left almost no bubbles in the clear wood coating's film [46]. Mildewcides, surfactants, defoamers, UV-stabilizers, wetting and dispersing agents are very common additives used in water-based wood coating formulations.

4. Influence of coating characteristics

4.1. Coating types

Wood coatings are divided into two major categories known as film forming and non-film forming or penetrating stains. Film-forming coatings for wood are mainly paints or varnishes (lacquers). The term paint is used for film-forming coatings that are highly pigmented and mask the substrate underneath completely. Paints are commonly used for wooden windows, doors and siding. They usually have the highest durability performance on wood, due to the fact that they contain a high concentration of pigments which protects the wood surface from UV degradation [47]. Varnishes and lacquers are clear film-forming formulations that are not designed for exterior application. Williams et al. [12] showed that clear varnishes will crack and peel from wood surfaces in less than 2 years.

Stains are capable of penetrating into the wood and were generally considered to not form any film layer on the surface. However, depending on the resin type and viscosity, different stains may also form a film layer on the surface of the wood [9]. **Figure 5** shows scanning electron microscope photos of cross sections of southern pine samples coated with two different semi-transparent penetrating stains with different viscosities. The coating with higher viscosity has much greater film thickness than the one with lower viscosity, and although these two formulations are designated as stains, they both form a film layer. Stains are divided in three different categories: transparent, semi-transparent, and solid colour stains. Transparent or clear stains either do not contain any pigment or contain nanopigments. Because clear stains do not have any pigment they need other types of UV stabilizer or blocker to be able to protect the wood from UV degradation. Wood coated with transparent or clear stains will usually turn to grey at the same rate as uncoated wood samples. As the wood below the coating degrades, the film will crack and then slowly peel from the surface. On the other hand, semi-transparent stains have some degree of pigment to protect the wood from UV-degradation while not too much to hide the beauty of wood grain. Solid colour stains were defined as penetrating formulations which usually provide the highest degree of protection by completely blocking the harmful effect of UV, yet they still do not form a thick film layer like paints; therefore, they are less likely to crack or peel from wood surface the way that varnishes and paints do.

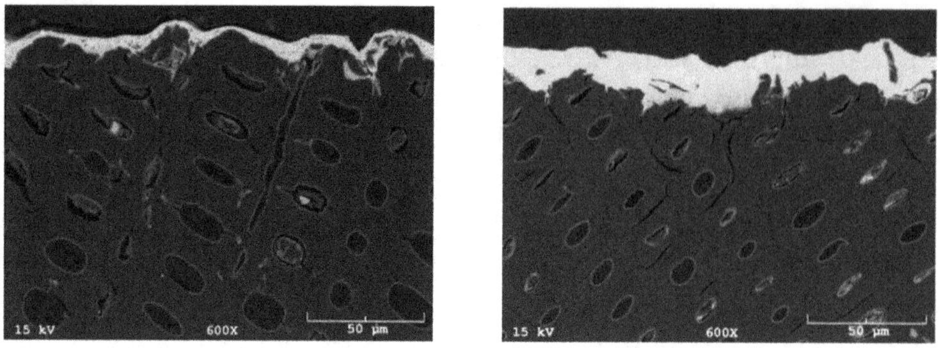

Figure 5. Back scattered electron images of cross-sections of pine samples coated with semi-transparent stains: alkyd-acrylic water-based stain (left) and a PU water-based formulation (right).

4.2. Coating properties

Understanding how different physical, chemical, and thermal properties of coatings affect their performance on woods in exterior conditions will help coating chemists develop more durable coatings. As the discussion of chemical properties of coating needs more in-depth overview of polymer and organic chemistry, this chapter will mostly be focused on the thermal and physical properties of coatings and their correlations with their performance on wood.

Testing wettability (the ability of liquid to spread on the substrate) of a coating is the first step in defining adhesion of a coating to wood or any other substrate. Measuring the contact angle of a droplet of coating on the substrate is used as a measure of wettability of the coating formulation. The lower the contact angle means better wetting and spreading of coating on wood. In general, to have good wetting, coatings should have much lower surface tensions compared to the surface energy of the substrate [48]. Wood has an average surface energy of 44 mN/m [49]. Solvent-based coatings have surface tensions of around 25 mN/m, whereas water-based coatings have on average surface tension of 32 mN/m [25]. Although both are lower than average surface energy of wood, reducing the surface tension of water-based systems will improve their adhesion and performance on wood.

The surface tension of water is 72 mN/m, and in water-based coatings, different additives like wetting agents and emulsifiers are added to the formulation to reduce the surface tension of water-based coating to be as close as possible to the surface tension of solvent-based coatings. To ensure the adhesion of a coating to wood, the contact angle test should be performed on late wood areas, because the contact angle of the coating is usually much higher on late wood than early wood [25]. This is another reason why erosion of coatings usually starts first from latewood sections then early wood and is most evident when flat-grained coated wood is exposed to weathering [50].

Another important coating property is glass transition temperature, which is an indicator of polymer flexibility and how it changes as it is heated. Amorphous polymers go through a transition from a glassy state to a flexible state in a temperature range defined as glass transition temperature (Tg). Previous researchers indicated that coatings formulated for exterior wood should penetrate deep into wood and not form any film layer on the top [12, 47].

They reported that film-forming coatings will crack and peel from the surface due to stresses caused by dimensional instability of wood when exposed outside [51]. This is entirely correct when the coating film is not flexible. If the dry coating film has adequate flexibility to tolerate stresses caused by swelling and shrinking of the wood in outdoor conditions, then the coating will be less likely to crack or peel from the surface. The flexibility of a coating film can be measured and defined by its glass transition temperature (Tg). If a coating has Tg lower than the temperature that wood-coated samples will be exposed to during its service life, then it will have high durability and good performance [25, 52]. It should be also noted that the glass transition of the coating may change in service. Podgorski [52] studied the changes in Tg of alkyd coating in natural weathering and reported that the Tg increases initially until it reaches a maximum value when the cross-linking is complete.

One of the basic physical properties of a coating which can be measured easily is viscosity. Viscosity plays a significant role in wood coating performance [25] and is highly correlated to coating film thickness. Lower viscosity coatings penetrate deeper into the wood, whereas higher viscosity coatings tend to form a film-layer on the surface [25]. Nussbaum et al. [53] studied the penetration of solvent-based alkyd versus water-based alkyd coatings and found that coatings with similar viscosities had similar penetration depth into the wood. Furthermore, we observed that higher viscosity coatings which also had higher film thickness and lower Tg (higher flexibility) performed better throughout 3 years of natural weathering of the wood tested in Canada [17].

4.3. Coating performance

Coatings have very complex formulations and their interaction with a biological material like wood makes prediction of their service life and performance much harder [54]. The average service life of exterior wood stains is between 1 and 3 years [55]. Coatings can reduce water uptake, UV-degradation, and checking of wood, which prolongs the service life of wooden structures; however, application of coating every year or two places the wooden products as high-maintenance products when compared to other building materials. Increasing the lifespan of coatings would reduce maintenance, thus encouraging homeowners, architects and contractors to use more wooden products.

There is no doubt that wood properties considerably affect the performance of coatings [6, 14]. The grain characteristics, surface texture, heartwood/sapwood, knots, extractives, moisture content and density are all factors that should be considered when predicting or evaluating the performance of coatings on wood [9, 50, 56].

Treating wood with preservative chemicals such as chromated copper arsenate (CCA) will change the surface property of wood, thus, affecting the coating's life span and durability. Previous studies reported better performance of coating on CCA-treated woods due to presence of chromium [57, 58]. CCA-treated wood has been reported to have lower water uptake and higher colour stability than untreated wood [17] which could be due to presence of chromium oxide in the CCA formulation that can modify the lignin [59]. As pressure treating wood with CCA solution did not change the wettability of the wood [60] while it did improve its water repellency and UV stability, formulating coating for CCA-treated wood was less challenging than any other treated wood.

Alkaline Cu-based preservative solutions such as ACQ and CA and micronized copper formulations that are arsenic and chromium free have now replaced CCA-treated wood in residential markets in North America. Wood treated with alkaline copper quaternary (ACQ) and copper azole (CA) has higher water uptake than even untreated wood [17]. However, they have much lower colour change than untreated wood when exposed to natural weathering [17]. The presence of copper oxide in these formulations has the potential to modify lignin in wood [20, 61]. As lignin is the most susceptible wood component to UV-degradation, modification of lignin in wood could be responsible for improved photo-stability of Cu-based treated wood [1].

Heat treatment is another wood modification technique which is becoming more popular as a green modification method. Heat treatment changes the chemical and physical properties of wood [33], thereby affecting coating performance and its adhesion to the modified wood. Researchers have reported contradictory results for changes in surface energy of thermally modified wood. Some found that heat treatment reduced the surface energy of wood [35, 62], and others reported improvements or no changes after modifications [63, 64]. A few studies also observed reduced adhesion of coating to heat-treated woods in comparison with untreated woods [36, 65]. However, what all researchers agree on is the lower water uptake of heat-treated due to the decrease in wood hygroscopicity after thermal modification [32, 33, 66]. Heat-treated wood has lower water uptake than preservative-treated wood but turns to grey at the same rate of untreated wood [34, 67, 68]. Therefore, coatings that are formulated for heat-treated wood should have higher UV resistance than coatings that were formulated previously for CCA-treated or other preservative-treated woods.

5. Conclusions

The key to designing a coating with better performance on wood is to consider not just the characteristics or interface of the coating and the wood themselves, but to study the interaction among individual wood constituents with the coating components. Close collaboration between wood scientists and coating chemists is the best way to develop new formulations with improved durability. Considering how wood components and surface properties of wood change when exposed to weathering, chemical, or thermal modification will help coating formulators to know where to concentrate research and development efforts. For instance, for heat-treated wood, a coating with higher UV resistance is required, whereas for Cu-based preservative, a coating with higher water repellency is beneficial. Additionally, a closer look at coating properties is critical in predicting the performance of new formulations. Measuring glass transition temperature (Tg) of new coating formulations and monitoring Tg changes during weathering is not a common practice by coating companies. However, this should be a standard practice for those who are formulating coatings for wood. A flexible coating with low Tg can better tolerate stresses caused by the dimensional instability of wood during swelling and shrinking. Thus, such a coating would be more resistant to cracking and peeling compared to a coating with higher Tg and rigid film. In conclusion, to develop exterior wood coatings with prolonged service lives, coating formulators should consider the following parameters: interactions of coating components with wood, coating properties and changes in the surface properties of wood after modification.

Author details

Mojgan Nejad[1]* and Paul Cooper[2]

*Address all correspondence to: Nejad@msu.edu

1 Michigan State University, East Lansing, Michigan, USA

2 University of Toronto, Toronto, Canada

References

[1] Hon, D.N.S. and Shiraishi N., Wood and Cellulosic Chemistry. 2001, New York: Marcel Dekker. 914.

[2] Feist, W.C. and Hon D.-S., Chemistry of weathering and protection. Advances in Chemistry Series, 1984(207): pp. 401–451.

[3] Hon, D.N. and Ifju G., Measuring penetration of light into wood by detection of photo-induced free radicals. Wood Science, 1978. **11**(2): pp. 118–127.

[4] Schniewind, A., Mechanism of check formation. Forest Products Journal, 1963. **13**(11): pp. 475–480.

[5] Evans, P.D., Urban K., and Chowdhury M.J.A., Surface checking of wood is increased by photodegration caused by ultraviolet and visible light. Wood Science and Technology, 2008. **42**: pp. 251–265.

[6] Forest Products Laboratory, ed. Wood Handbook—Wood as an Engineering Material. 1999, Madison, WI: U.S. Department of Agriculture, Forest Service, Forest Products Laboratory. 463.

[7] Kahn, M., Fine Woodworking on Wood and How to Dry it, ed. F.w. working. 1986, United States: Taunton Press. 106.

[8] Feist, W.C., Painting and finishing exterior wood. Journal of Coatings Technology, 1996. **68**(856): pp. 23–26.

[9] Williams, R.S., et al., Wood properties affecting finish service life. Journal of Coatings Technology, 2000. **72**(902): pp. 35–42.

[10] Simpson, W. and TenWolde A., Physical properties and moisture relations of wood. Chapter, 2010. **3**: pp. 1–25.

[11] Van Den Bulcke, J., et al., Adhesion and weathering performance of waterborne coatings applied to different temperate and tropical wood species. Journal of Coatings Technology Research, 2006. **3**(3): pp. 185–191.

[12] Williams, R.S., Knaebe M.T., and Feist W.C., Finishes for Exterior Wood. 1996, Madison, WI, USA: Forest Products Laboratory. 127.

[13] Palanti, S., et al., A simple testing method for the measurement of the water vapour transmission of coated wood longitudinal and tangential to grain direction. Holzforschung, 2001. **55**: p. 7.12.2005.

[14] Williams, R.S., et al., Wood properties affecting finish service life. Journal of Coatings Technology, 2000. **72**(902).

[15] Bonura, T., D. Mall, and S. Williams, Finishing checklist-A guide to achieving optimum coating performance on exterior wood surfaces. Journal of Coatings Technology, 2004.

[16] Lebow, S.T., Wood Preservation, in Wood Handbook—Wood as an Engineering Material, F.P. Laboratory, Editor. 2010, Washington: U.S. Department of Agriculture: p. 28.

[17] Nejad, M. and Cooper, P., Exterior wood coatings. *Part*-1: Performance of semitransparent stains on preservative-treated wood. Journal of Coatings Technology Research, 2011. **8**(4): pp. 449–458.

[18] Freeman, M.H., et al., Past, present, and future of the wood preservation industry. Forest Products Journal, 2003. **53**(10): pp. 8–15.

[19] Feist, W.C. and Ross A.S., Performance and durability of finishes on previously coated CCA-treated wood. Forest Products Journal, 1995. **45**(9): pp. 29–36.

[20] Evans, P.D., Michell A.J., and Schmalzl K.J., Studies of the degradation and protection of wood surfaces. Wood Science and Technology, 1992. **26**(2): pp. 151–163.

[21] Lebow, S., Leaching of wood preservative components and their mobility in the environment: summary of pertinent literature, in General technical report F.P. Laboratory, Editor. 1996: Forest Products Laboratory. p. 36.

[22] Lebow, S., P. Lebow, and D. Foster, Estimating preservative release from treated wood exposed to precipitation. Wood and Fiber Science, 2008. **40**(4): pp. 562–571.

[23] Cooper, P.A. and Ung Y.T., Effect of preservative type and natural weathering on preservative gradients in southern pine lumber. Wood and Fiber Science, 2009. **41**(3): pp. 229–235.

[24] Nejad, M. and Cooper P., Coatings to reduce wood preservative leaching. Environmental Science and Technology, 2010. **44**(16): pp. 6162–6166.

[25] Nejad, M. and Cooper P., Exterior wood coatings. Part-2: Modeling correlation between coating properties and their weathering performance. Journal of Coatings Technology Research, 2011. **8**(4): pp. 459–467.

[26] Rowell, R.M., et al., Understanding decay resistance, dimensional stability and strength changes in heat-treated and acetylated wood. Wood Material Science and Engineering, 2009. **4**(1–2): pp. 14–22.

[27] Hill, C.A.S., Wood Modification: Chemical, Thermal and Other Processes Wiley series in renewable resources. 2006, NJ: John Wiley & Sons, Ltd.

[28] Alfredsen, G., Flæte P.O., and Militz H., Decay resistance of acetic anhydride modified wood: a review. International Wood Products Journal, 2013. **4**(3): pp. 137–143.

[29] Gobakken, L.R. and Westin M., Surface mould growth on five modified wood substrates coated with three different coating systems when exposed outdoors. International Biodeterioration and Biodegradation, 2008. **62**(4): pp. 397–402.

[30] Ohkoshi, M., FTIR-PAS study of light-induced changes in the surface of acetylated or polyethylene glycol-impregnated wood. Wood Science, 2002. **48**(5): pp. 394–401.

[31] Beckers, E.P.J., et al., Performance of finishes on wood that is chemically modified by acetylation. Journal of Coatings Technology, 1998. **70**(878): pp. 59–67.

[32] Wang, J. and P. Cooper. Review on thermal treatments of wood. in CWPA Proceedings. 2003. Halifax, Canada.

[33] Esteves, B.M. and Pereira H.M., Wood modification by heat treatment: A review. BioResources, 2009. **4**(1): pp. 370–404.

[34] Dubey, M.K., Pang S., and Walker J., Color and dimensional stability of oil heat-treated radiata pinewood after accelerated UV weathering. Forest Products Journal, 2010. **60**(5): pp. 453–459.

[35] Pétrissans, M., et al., Wettability of heat-treated wood. Holzforschung, 2003. **57**(3): pp. 301–307.

[36] Nejad, M., et al., Coating performance on oil-heat treated wood for flooring. Bioresources 2012. **8**(2): pp. 1881–1892.

[37] Zeno W. Wicks Jr., et al., Organic Coatings: Science and Technology. 2007, Hoboken, N.J.: Wiley-Interscience.

[38] Winkelaar, A., Coatings Basics ed. E.C.T. Files. 2010: William Andrew, Hannover, Germany. 140.

[39] Bently, J. and Turner G.P.A., Introduction to Paint Chemistry and Principle of Paint Technology. 1998, London: Chapman and Hall.

[40] Bulian, F. and Graystone J.A., Industrial Wood Coatings-Theory and Practice. 1st ed. 2009, Oxford, UK: Elesvier.

[41] Weldon, D.G., Failure Analysis of Paints and Coatings. 2002: John Wiley & Sons, West Sussex, United Kingdom. 291.

[42] Jeremic, D., Cooper P., and Brodersen P., Penetration of poly(ethylene glycol) into wood cell walls of red pine. Holzforschung, 2007. **61**(3): pp. 272–278.

[43] Landry, V., Riedl B., and Blanchet P., Alumina and zirconia acrylate nanocomposites coatings for wood flooring: Photocalorimetric characterization. Progress in Organic Coatings, 2008. **61**(1): pp. 76–82.

[44] Nejad, M., et al., Studying dispersion quality of nanoparticles into a bio-based coating. Progress in Organic Coatings, 2015. **89**: pp. 246–251.

[45] Heilen, W., Haim J., and Hyatt D., Additives for Waterborne Coatings. 2009: Vincentz Network, Hannover, Germany.

[46] Nejad, M., et al., Waterborne coating: defoamer performance measured by evaluation of coating film appearance. European Coating Journal, 2015. (10): p. 6.

[47] Satas, D. and Tracton A.A., Coatings Technology Handbook. 2001, New York: Marcel Dekker. 902.

[48] Packham, D.E., ed. Handbook of Adhesion. 2nd ed. 2005, John Wiley: Chichester, West Sussex. 639.

[49] Wålinder, M., Wetting Phenomena on Wood: Factors Influencing Meaurements of Wood Wettability in Department of Manufacturing Systems, Wood Technology and Processing. 2000, Sweden: Royal Institue of Technology: KTH.

[50] Feist, W.C., Wood properties and finish durability. Journal of Coatings Technology, 2002. **74**(926): pp. 71–76.

[51] Cassens, D.L. and Feist W.C., Exterior Wood in the South, Selection, Applications, and Finishes, in General Technical Report, F.P. Laboratory, Editor. 1991: Madison, WI: US department of Agriculture. p. 60.

[52] Podgorski, L.. Analysis of the wood coating ageing and prediction of the durability through calorimetric investigations. in COST E 18 Final seminar 6. 2004.

[53] Nussbaum, R.M., Sutcliffe E.J., and Hellgren A.C., Microautoradiographic studies of the penetration of alkyd, alkyd emulsion and linseed oil coatings into wood. Journal of Coatings Technology, 1998. **70**(878): pp. 49–57.

[54] Dickie, R.A., Toward a unified strategy of service life protection. Journal of Coatings Technology, 1992. **64**(809): pp. 61–64.

[55] Knaebe, M., Paint, Stain, Varnish, or Preservative? It's your Choice., in The finish line, fact sheet. 1995, Madison, WI Forest products laboratory.

[56] Jourdain, C., et al., Changing nature of wood products–what does it mean for coatings and finish performance? (Technology Forum: Wood Coatings), 1999. **71**(890): pp. 61–66.

[57] Feist, W. and Williams R.S., Weathering durability of chromium-treated southern pine. Forest Products Journal, 1991. **41**(1): pp. 8–14.

[58] Feist, W. and Ross A.S.. Performance of surface finishes over CCA-treated wood. In 43 rd annual meeting 1989. Madison: Forest Products Reaserch Society.

[59] Michell, A., FTIR spectroscopic studies of the reactions of wood and of lignin model compounds with inorganic agents. Wood Science and Technology, 1992. **27**(1): pp. 69–80.

[60] Ross, A., et al., Finishability of CCA Pressure-Treated Wood, in Miscellaneous Publication, P.a.C. Industry, Editor. 2000, Madison, WI: USDA, Forest Products Laboratory. pp. 44–58.

[61] Grelier, S., Castellan A., and Kamdem D.P., Photoprotection of copper-amine-treated pine. Wood and Fiber Science, 2000. **32**(2): pp. 196–202.

[62] Kocaefe, D., et al., Effect of heat treatment on the wettability of white ash and soft maple by water. Einfluss der Wärmebehandlung auf die Benetzbarkeit von Weißesche und Rot-Ahorn mit Wasser, 2008. **66**(5): pp. 355–361.

[63] Wolkenhauer, A., et al., Plasma treatment of heat treated beech wood–investigation on surface free energy. Holzforschung, 2008. **62**(4): pp. 472–474.

[64] Awoyemi, L., Cooper P., and Ung T., In-treatment cooling during thermal modification of wood in soy oil medium: soy oil uptake, wettability, water uptake and swelling properties. European Journal of Wood and Wood Products, 2009. **67**(4): pp. 465–470.

[65] PetriÄ M., et al., Wettability of waterborne coatings on chemically and thermally modified pine wood. Journal of Coatings Technology Research, 2007. **4**(2): pp. 203–206.

[66] Sailer, M., Rapp A.O., and Leithoff H.. Improved resistance of scots pine and spruce by application of an oil-heat treatment. in The International Research Group on Wood Preservation. 2000. Hawaii, USA: IRG/WP 00-40162.

[67] Nejad, M. and Cooper P., Performance characterization of coatings on treated-wood. 2013, Progress in Color, Colornats and Coatings. pp. 61–65.

[68] Jämsä, S., Ahola P., and Viitaniemi P., Long-term natural weathering of coated ThermoWood. Pigment & Resin Technology, 2000. **29**(2): pp. 68–74.

11

Flame-Retardant Systems Based on Alkoxysilanes for Wood Protection

Carlos A. Giudice and Guadalupe Canosa

Abstract

The aim of this study was to formulate, develop, and determine the performance of flame-retardant systems for wood protection. Flame-retardant systems involve wood impregnation and intumescent coating application. The impregnation was made in two retention levels using silanes of low and high hydrophobicity (methyltriethoxysilane and n-octyltriethoxysilane, respectively); these silanes were conducted to polymerize by sol-gel process in wood pores. The intumescent coatings were formulated with a polymeric binder (hydroxy-functional acrylic resin) modified with n-octadecyltriethoxysilane in two w/w ratios to combine the individual characteristics of each film-forming material. In this research, *Pinus radiata* panels were selected to study the performance of quoted flame-retardant systems. The results indicated excellent flame-retardant performance of some studied systems in two foot tunnel, in oxygen index cabin, and in horizontal-vertical chamber.

Keywords: wood, alkoxysilane, impregnation, intumescent coating, fire performance

1. Introduction

The use of wood in construction is often questioned by its natural combustibility and vulnerability in fires [1–3]. Nevertheless, the excellent resistance to fire penetration due to the low thermal conductivity and the ability to form a superficial char layer allow wood to maintain the physical-mechanical properties longer than metal or concrete [4–6].

The protection of wood against fire is limited to a retarding effect, since no chemical can transform wood in a noncombustible material within reasonable economic margins. Flame-retardants are applied to significantly reduce or delay the combustion of a material. Nevertheless, it is worth mentioning that the above-mentioned wood treatments only control small sources of energy (heat and/or fire) prolonging the beginning of conflagration and so providing the time to take the necessary actions [7–11].

Experiences and investigations on flame-retardant products show a wide range of substances that prove positive action, for example, on reducing the mass loss, the flame advance, and the permanence time of combustion and incandescence; these methods to improve fire performance of wood include chemical and/or physical modifications, as well as structural design considerations [12, 13].

High efficient flame-retardant systems for wood protection usually consist of impregnation and/or intumescent coating application. Several techniques have been studied for impregnations (chemicals in molecular dispersions and colloidal solutions) and for surface treatments (flame-retardant coatings), both of them with cutting-edge technologies (e.g., nanocomposites) [1]. The main problem of these flame-retardant treatments is the long-term performance in weather exposure where flame-retardant chemicals may leach out; while in indoor, the most important issue is the persistence of the aesthetic appearance that cannot always predict or ensure [12]. The use of intumescent coating is one of the most economical and efficient ways to protect materials against fire action without modifying their intrinsic properties [14–16] as the impregnation does [17, 18].

In previous papers, the authors studied the performance of various flame-retardant systems for wood protection; experiments on treated woods allowed to conclude that the performance of the impregnations depend on whether they give chemical or physical changes or both simultaneously [19–21] while that the efficiency of the intumescent coatings is strongly influenced by type (chemical composition) and content (pigment volume concentration, PVC) of functional pigments as well as by dry film thickness [22, 23].

Nowadays, many organosilicon compounds are chosen as environmental-friendly fire retardants to replace the traditional halogenated ones, for several materials protection including the wood; the most commonly used compounds include silanes, polysiloxanes, and polysilsesquioxanes, as well as structures containing heteroatoms in the main chain, such as polycarbosilane and polysilazanes [24, 25].

The aim of this research was to design highly efficient flame-retardants for the protection of *Pinus radiata* panels; the studied systems included an impregnation based on low and high hydrophobicity silanes (methyltriethoxysilane (MTES) and n-octyltriethoxysilane (OTES), respectively) polymerized inside wood pores by sol-gel process, and the subsequent application of a coating with intumescent properties formulated with a polymeric binder (hydroxy-functional acrylic resin) chemically modified with n-octadecyltriethoxysilane (ODTES) to improve the individual characteristics of each film-forming material.

2. Materials and Methods

2.1. Impregnation

The experiments included: (i) the preparation of wood panels; (ii) the choice of reactive impregnants; (iii) the selection of impregnation conditions; (iv) the setting of drying and curing conditions; and (v) the determination of impregnant retention.

Preparation of wood panels. Pinus radiata was selected because it is a macroporous and moderately penetrable wood. Boards from unseasoned trunks were provided by a sawmill; they were exposed in a chamber under controlled temperature and humidity conditions (20 ± 2°C and 60 ± 5% RH) until reaching the equilibrium moisture, according to the guidelines of ASTM D4933. Free-defect specimens were prepared, with the appropriate size for each test, using a power saw; then, the specimens were carefully sanded with fine sandpaper.

The cellulose was activated by immersion in a sodium hydroxide solution (pH 8.5); then, the specimens were washed with distilled water to remove the remaining alkali [26, 27]. Finally, the samples were again conditioned to reach the equilibrium moisture and sanded.

Reactive impregnants. Since silicon-derived monomers with low molecular weight react with the cellulose -OH groups present in the internal wall of wood pores [7], methyltriethoxysilane (MTES) and n-octyltriethoxysilane (OTES) were selected to obtain impregnants with different hydrophobicity (**Table 1**). For the impregnation, the silanes were solubilized in 1/1 v/v toluene/ethylenglycol (concentration, 12% v/v).

Chemical name	Methyl triethoxysilane	n-octyl triethoxysilane	n-octadecyl triethoxysilane
Chemical structure	$(CH_3)\text{-}Si\text{-}(OC_2H_5)_3$	$C_8H_{17}\text{-}Si\text{-}(OC_2H_5)_3$	$C_{18}H_{37}\text{-}Si\text{-}(OC_2H_5)_3$
Empirical formula	$C_7H_{18}O_3Si$	$C_{14}H_{32}O_3Si$	$C_{24}H_{52}O_3Si$
Abbreviation	MTES	OTES	ODTES
Aspect	Colorless liquid	Colorless liquid	Colorless liquid
Molecular weight	178.30	276.48	416.76
Density (25°C), g/cm³	0.895	0.880	0.875
Purity (gas chromatography), %	99.0	98.0	97.0

Table 1. Silanes properties.

Impregnation conditions. The impregnation was carried out under controlled operating conditions at 48–50°C in an autoclave equipped with vacuum pump and compressor. The autoclave was initially charged with the wood panels. Then, it was applied 500 mm Hg vacuum for 10 min to remove air and water vapor from pores to facilitate the penetration of chemical modifier; later, the solution of corresponding alkoxide was incorporated without decreasing the vacuum level. In all cases, the impregnating solution/wood ratio was 3/1 v/v to ensure that the panels were completely submerged during the entire process.

The operating conditions were adjusted to obtain two groups of panels with different retention levels of chemical modifier. To facilitate the penetration, the pressure was gradually increased from 1.5 to 6.5 kg/cm^2; this stage lasted between 15 and 60 min. Subsequently, a slight vacuum (about 200 mm Hg for 10 min) was applied to remove the excess of solution from panel surface.

Drying and curing conditions. After impregnation, wood samples were taken off from the autoclave, rinsed and exposed to controlled laboratory conditions (20 ± 2°C and 60 ± 5% RH) for 2 days to allow the complete drying (removal of organic solvent), the partial development of the curing reactions involved in the sol-gel process (hydrolysis, water absorption from the environment, and formation of water and ethylic alcohol by condensation), and, finally, the elimination of quoted products to the surrounding (aging). This process produces a chemical modification of wood, forming a polymer coating in the internal wall of the pores without sealing off [28–33].

Impregnant retention. It was gravimetrically measured and, as mentioned, two panel groups were selected: 45–50 and 85–90 kg/m^3. Finally, it was microscopically observed that the depth reached by the impregnants was complete (entire thickness of the panel) in the highest retention level and only partial in the other.

2.2. Hybrid intumescent coatings

The experimental part included: (i) the formulation; (ii) the manufacture; and finally (iii) the application.

Formulation. The intumescent hybrid coatings were based on a polymeric material chemically modified with a long chain silane (**Table 2**).

Component	% v/v on dry film
Ammonium polyphosphate	48.3
Pentaerythritol	17.7
Melamine	17.1
Titanium dioxide	6.9
Hydrated zinc borate	2.0
Alumina trihydrate	1.8
Additives	6.2
Film forming material (solids)	100.0
PVC = 65.0%	

Table 2. Intumescent coatings composition.

Film-forming material. A hydroxy-functional acrylic resin was used (**Table 3**). The hybrid binders included this resin and the n-octadecyltriethoxysilane (ODTES, **Table 1**). One of them was formulated with resin/ODTES in stoichiometric ratio (2.5/1.0 w/w): it was contemplated the three hydroxyl groups of silanol (generated by hydrolysis of triethoxysilane, 12.2% w/w) and the percentage of hydroxyl groups of the resin (4.9% w/w). The other was formulated with an excess amount of ODTES with respect to stoichiometric ratio (1.0/1.0 w/w). In addition, a nonreactive (thermoplastic) acrylic resin based on methyl methacrylate diluted in xylene was selected as reference (**Table 4**).

Solid content, %	60.2
Solvent	Xylene/ethylenglycol acetate, 2/1 v/v
Viscosity (25°C)	X-Z Gardner; 15.2 Stokes
Acid value, mg KOH/g	6.1
Density (25°C), g/cm³	0.972
HO^- on solids, %	4.9

Table 3. Properties of the hydroxy-functional acrylic resin.

Solid content, %	59.8
Solvent	Xylene
Viscosity (25°C)	Z2-Z4 Gardner
Acid value, mg KOH/g	4.0
Density (25°C), g./cm³	0.825

Table 4. Properties of the nonreactive acrylic resin.

Pigmentation. The active pigments were the dipentaerythritol (water solubility at 20°C, 0.29 g/100 ml) as carbon source, an ammonium polyphosphate (white powder with 30.2% phosphorus and 14.5% nitrogen and low water solubility at 20°C, 0.18 g/100 ml) as acid catalyst, and, finally, a modified melamine (decomposition temperature about 130°C; water solubility at 20°C, 0.32 g/100 ml) as blowing agent [22, 34].

Complementary flame-retardant pigments were a hydrated zinc borate ($2ZnO.3B_2O_3.7.5H_2O$) and the alumina trihydrate ($Al_2O_3.3H_2O$) [30–32]. Rutile titanium dioxide (TiO_2) was used as opaque pigment (**Table 5**).

Other components. Rheological, dispersing, and stabilizing agents were selected as additives. The solvent mixture for film-forming materials was 2/1 v/v toluene/ethylenglycol acetate.

Manufacture of intumescent coatings. It was carried out in a high-speed disperser. In a first stage, the solvent mixture was added; subsequently, the rheological agent was gradually introduced (castor oil gel, 15% w/w) at high shear rate (1400 rpm), until reaching a system with the suitable viscosity for pigment dispersion (laminar flow, rolling-doughnut effect); the stirring was extended for about 10 min. Then, surfactants (dispersants and dispersion stabilizers) and pigments were incorporated under vigorous stirring for about 30 min to achieve adequate dispersion. Finally, at low shear rate (about 700 rpm), the hydroxy-functional acrylic resin was incorporated to achieve 65% pigment volume concentration (PVC). The corresponding silane solution (50% v/v) was placed in a second container in sufficient amount to achieve the 2.5/1.0

and 1.0/1.0 w/w ratios depending on the formulation considered (expressed in solids of acrylic resin/silane).

Alumina trihydrate	Appearance, granular crystalline powder Density, 2.4 g/cm^3 Oil absorption, 53 g/100 g Particle diameter D (50/50), 0.65 μm
Hydrated and stabilized zinc borate	Appearance, white powder Density, 3.9 g/cm^3 Oil absorption, 28 g/100 g Particle diameter D (50/50), 5.20 μm
Titanium dioxide (rutile)	Appearance, white powder Density, 4.1 g/cm^3 Oil absorption, 39 g/100 g Particle diameter D (50/50), 0.25 μm

Table 5. Characteristics of pigments.

In the case of reference coating, after finishing pigment dispersion, the nonreactive acrylic resin was incorporated in the same container (also 65% PVC).

Application of intumescent coatings. First, the content of the two packages was mixed in adequate amounts; in all samples the volume solids content, using the quoted solvent mixture, was adjusted to 50%. Subsequently, the impregnated panels (dried and partially cured) were coated with the intumescent products by brush (two layers, with an interval of 24 h).

In all cases, and to ensure the drying and curing of protective treatments, the panels were kept in controlled laboratory conditions (25 ± 2°C and 65 ± 5% RH) for 7 days. Dry film thickness ranged between 160 and 180 μm.

Visual and microscopic observations indicated the absence of checking and cracking, typical of the films based on only silanes as film-forming material. The identification of the panels is displayed in **Table 6**.

2.3. Fire performance

Two foot tunnel. Flame advance (AL) was determined by using equation AL = Ls − La, where Ls is the average of the three highest flame advances on the panels (consecutive readings measured at intervals of 15 s) and La is the average distance of flame reflection (55 mm) on a thin cement substrate selected as reference; panel consumption (PC) was also evaluated. Specimens were prepared with 6 × 100 × 605 mm size (radial × tangential × longitudinal); the tests were performed in triplicate according the guidelines of ASTM D3806. Then, the results were averaged.

Oxygen index, OI. This test determines the minimum oxygen concentration in mixture with nitrogen that can support the combustion of a material, under equilibrium conditions, as candle-like burning; the importance of OI determination lies in the high reproducibility of results. The test was performed in triplicate under the guidelines of ASTM D2863, at 23 ± 2°C and with a 3.2 cm/s flow rate; test specimen dimensions were 150 mm length, 10 mm width, and 10 mm thickness (longitudinal, radial, and tangential, respectively).

Resistance to intermittent flame of Bunsen burner, RIB. For the first stage, the panels were exposed in a horizontal-vertical cabinet for 20 s to the action of the flame with 10 s rest: the number of cycles with self-extinguishing behavior was defined in 30 as maximum (it was assigned one point per cycle); for the second stage, the flame action was extended to 50 s with 10 s rest: the number of cycles with self-extinguishing behavior was defined in 35 as maximum (it was assigned two points per cycle), and, finally, when the system continued behaving as self-extinguishing, the flame was kept constantly for 30 min as maximum (it was assigned 5 points per minute). Lastly, the total score for each panel was calculated (the highest rating can reach 250 points). Test specimen dimensions were 150 mm length, 40 mm width, and 10 mm thickness (longitudinal, radial, and tangential, respectively).

Impregnant		Coating	Identification
Type	Retention, kg/m^3		
–	–	Nonreactive acrylic	A.1
	–	Acrylic/ODTES (2.5/1.0 w/w ratio)	A.2
	–	Acrylic/ODTES (1.0/1.0 w/w ratio)	A.3
MTES	45–50	Nonreactive acrylic	A.4
	45–50	Acrylic/ODTES (2.5/1.0 w/w ratio)	A.5
	45–50	Acrylic/ODTES (1.0/1.0 w/w ratio)	A.6
	85–90	Nonreactive acrylic	A.7
	85–90	Acrylic/ODTES (2.5/1.0 w/w ratio)	A.8
	85–90	Acrylic/ODTES (1.0/1.0 w/w ratio)	A.9
OTES	45–50	Nonreactive acrylic	A.10
	45–50	Acrylic/ODTES (2.5/1.0 w/w ratio)	A.11
	45–50	Acrylic/ODTES (1.0/1.0 w/w ratio)	A.12
	85–90	Nonreactive acrylic	A.13
	85–90	Acrylic/ODTES (2.5/1.0 w/w ratio)	A.14
	85–90	Acrylic/ODTES (1.0/1.0 w/w ratio)	A.15
Without impregnant	–	–	Reference

Table 6. Panel identification.

3. Results

During experiences, visual and microscopic observations of treated panels allowed the description of the protective mechanism of flame-retardant systems: (i) when hybrid film is heated, the first resin layer is soften and the gases, produced by thermal decomposition of gas generator, are released allowing the film intumescence; (ii) then, the heat penetrates in the adjacent deeper layer where the inorganic components melt or soften retarding the heat conduction, while organics degrade into smaller products that contribute to formation of the carbonaceous residue (char); (iii) the heat continues penetrating and reaches an even deeper layer, causing the degradation and forming products that are transferred to the combustion zone through the char: the kinetics of the process seems to decrease as the system forms a carbonaceous material with high inorganic content that accumulates in the interfaces during combustion, isolates the underlying material, and reduces the mass loss; (iv) when the heat finally reaches the impregnated substrate, in the first instance it forms a carbonaceous residue by degradation until reaching the phase conformed by siloxanes, which acts as flame-retardant due to their high thermal stability; and finally, (v) this process is repeated generating interfaces of carbonaceous material until siloxanes reappear and the process starts again, giving to wood self-extinguishing characteristics.

Regarding test results, they are shown in **Table** 7. With the purpose of establishing the efficiency against the fire action of each protective system, the values 0 and 10 were set for 16 and ≥45% in OI; for 555 and 0 mm in AL; for 7.38 and 0.00% in PC; and finally, for 0 and 250 points in RIB. In all the cases, intermediate values were proportionally calculated. Finally, the values were averaged for the interpretation of the results: the highest average value indicates the best performance (**Table 8**).

Sample	Fire performance			
	AL, mm	PC, %	OI, %	RIB
A.1	165	2.22	31	175
A.2	134	1.80	36	192
A.3	138	1.89	33	188
A.4	60	1.23	40	204
A.5	49	1.11	45	214
A.6	52	1.18	43	210
A.7	38	0.52	42	228
A.8	32	0.35	45	236
A.9	33	0.41	44	230
A.10	58	1.33	37	200
A.11	52	1.20	40	210
A.12	54	1.23	38	205
A.13	40	0.58	40	220
A.14	28	0.38	45	233
A.15	35	0.44	43	225
Reference	555	7.38	16	2

Table 7. Test results.

Sample	Individual value	Average value
A.1	6.55	7.02
A.2	7.43	
A.3	7.08	
A.4	8.42	8.75
A.5	9.04	
A.6	8.79	
A.7	9.17	9.40
A.8	9.60	
A.9	9.43	
A.10	8.10	8.30
A.11	8.53	
A.12	8.29	
A.13	8.89	9.25
A.14	9.58	
A.15	9.27	
Reference	0.02	0.02

Table 8. Average flame-retardant performance of protective systems.

In relation to impregnant type, the analysis of the results indicates that MTES showed a higher performance than OTES (mean values, 9.08 and 8.78, respectively, **Figure 1**). This is based on that although MTES has a shorter hydrocarbon chain and therefore it is less able to generate the desired char, it forms siloxanes with higher silicon level (more percentage content of inorganic components) than OTES.

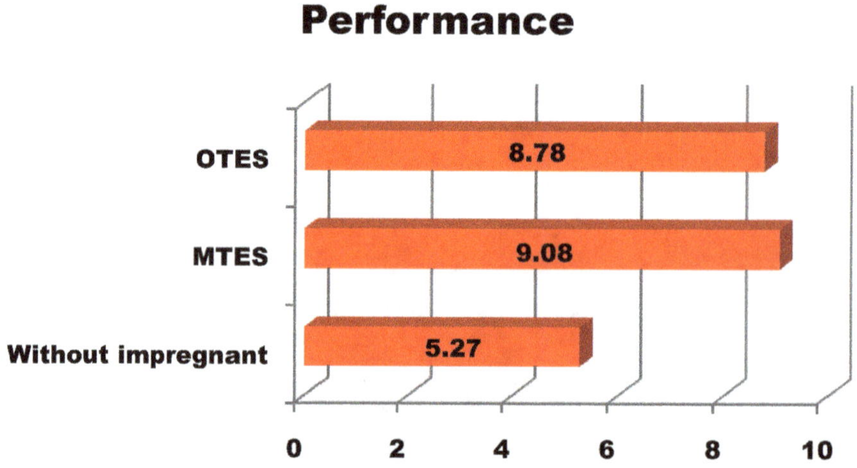

Figure 1. Flame-retardant performance according to the impregnant type.

With reference to retention level, flame-retardant efficiency improved as reactive impregnant level increased (mean values, 9.32 and 8.53 for 85–90 and 45–50 kg/m^3 respectively, **Figure 2**).

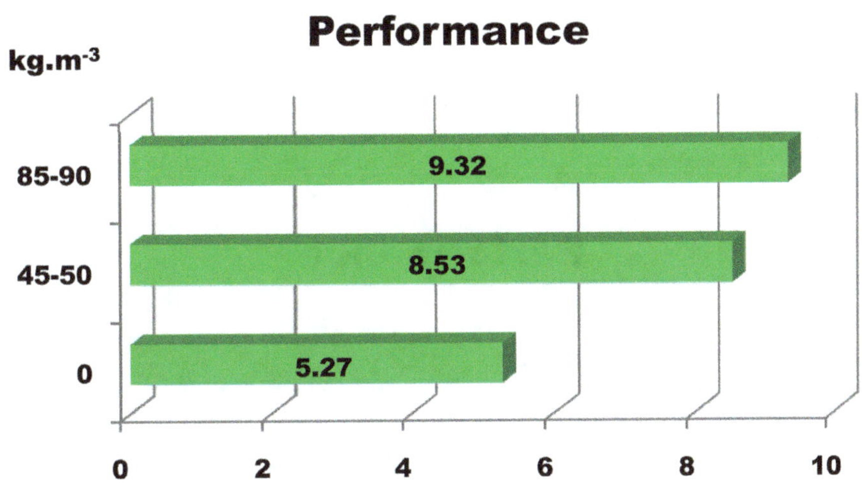

Figure 2. Flame-retardant performance according to the impregnant retention.

Finally, it is worth mentioning that panels treated with reactive impregnants displayed better performance with respect to those untreated (5.27). This would be supported by the chemical modification of wood produced by the formation of highly stable condensation products during the sol-gel process (≡Si-O-cellulose); the reactions involve the silanol groups generated by hydrolysis of the triethoxysilanes and the hydroxyl groups of wood cell wall.

In relation to wood panels also protected with hybrid intumescent coatings, it was observed that all impregnated samples displayed better performance against fire action than those without impregnation (**Table 8**). The above-mentioned is based on that chemical reactions in nonimpregnated panels would involve the hydroxyl groups of the film-forming materials and those of the wood cellulose, while in impregnated and partially cured panels, would react in addition to those groups that correspond to the impregnant. In the latter case, the highest thermal stability, attributable to the high quantity of strongly stable chemical bonds, complements the own performance of impregnants and of intumescent coatings.

Moreover, the tests allowed establishing that hybrid coatings were more efficient than those formulated with only one film-forming material and, in addition, that the hybrid coating formulated with hydroxy-functional acrylic resin/ODTES in the stoichiometric ratio showed the best performance (**Figure 3**).

SEM micrographs of chars corresponding to A.1, A.2, and A.3 panels are displayed in **Figure 4**. The analysis carried out on all panels indicates that: (i) the coating based on non-reactive acrylic resin had a moderately uniform distribution of cell size with some cracks; (ii) the coating modified with ODTES in the stoichiometric ratio generated a very uniform distribution of cell size without any crack; and finally, (iii) the coating modified with ODTES in excess led to a wide distribution of cell size with some cracks.

Above-mentioned indicates that the performance of treated panels depends strongly on the physical structure of the char formed by fire action. Correlating results of fire tests and SEM micrographs, it is observed that the best performance was achieved in panels with uniform chars free of cracks since they provide better insulation to the substrate.

In summary, the best flame-retardant system was that impregnated with MTES in the highest level of retention and superficially treated with the intumescent coating based on hydroxy-functional acrylic resin modified with ODTES in the stoichiometric ratio (**Table 8**; sample A. 8: average value, 9.60).

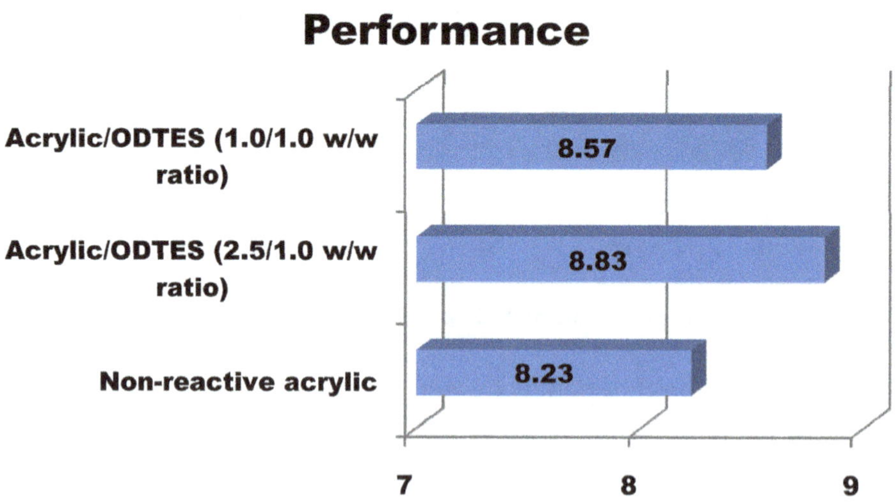

Figure 3. Flame-retardant performance according to the intumescent coating.

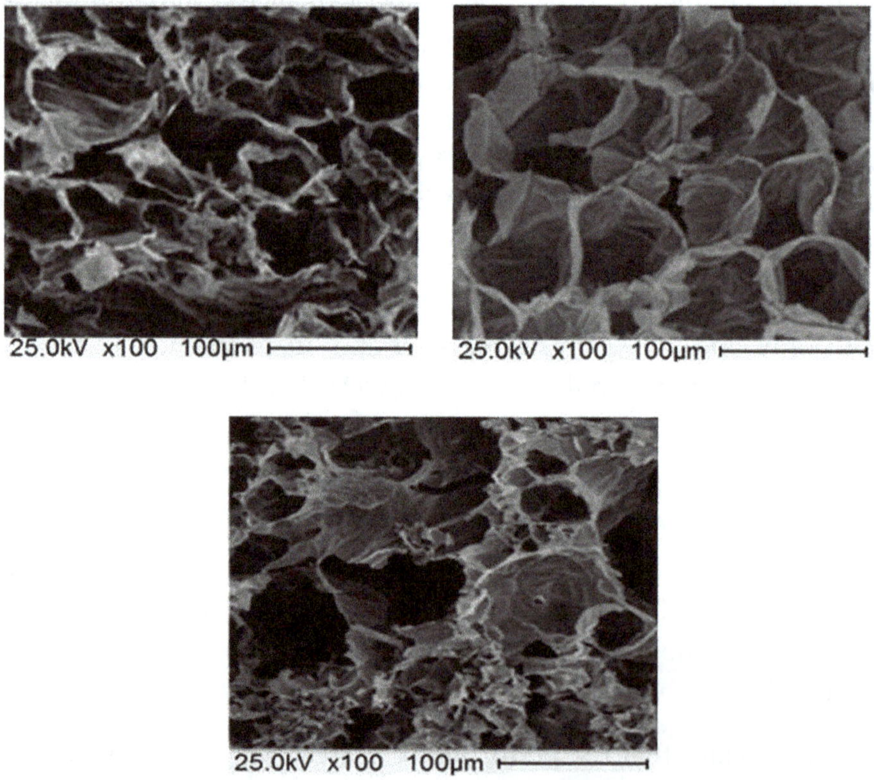

Figure 4. Char SEM micrographs: left, non-reactive acrylic resin; right, hydroxy-functional acrylic resin modified with ODTES in 2.5/1.0 w/w ratio and down, hydroxy-functional acrylic resin modified with ODTES in 1.0/1.0 w/w ratio.

Acknowledgements

The authors thank to National Technological University (UTN), to Scientific Research Commission of Buenos Aires Province (CICPBA), and to National Council of Scientific and Technical Research (CONICET) from Argentine for their sponsorship for conducting this research.

Author details

Carlos A. Giudice[1*] and Guadalupe Canosa[1,2]

*Address all correspondence to: cagiudice@yahoo.com

1 UTN-FRLP (National Technological University-Faculty Regional La Plata), La Plata, Argentine

2 CIDEPINT (Research and Development Center in Coatings Technology), La Plata, Argentine

References

[1] Bergman R, Cai Z, Carll C, et al. Wood handbook: wood as an engineering material. United States Department of Agriculture Forest Service, Forest Products Laboratory, Madison, Wisconsin; 2010. 508 p. General Technical Report FPL-GTR-190

[2] Athey R, Shaw P. The problem with fire. European Coatings Journal. 1999; 10:428–431.

[3] Koslowsky R, Przybylak M. Natural polymers, wood and lignocellulosic materials. In: Fire Retardant Materials, Horrocks and Price editors, CRC Press, UK; 2001. pp. 293–317. DOI: 10.1533/9781855737464.293

[4] Septien S, Valin S, Peyrot M, et al. Characterization of char and soot from millimetric wood particles pyrolysis in a drop tube reactor between 800 °C and 1400 °C. Fuel. 2014; 121:216–224. DOI: 10.1016/j.fuel.2013.12.026

[5] Inari G, Mounguengui S, Dumarçay S, et al. Evidence of char formation during wood heat treatment by mild pyrolysis. Polymer Degradation and Stability. 2007; 92(6):997–1002. DOI: 10.1016/j.polymdegradstab.2007.03.003

[6] Jiang T, Feng X, Wang Q, et al. Fire performance of oak wood modified with N-methylol resin and methylolated guanylurea phosphate/boric acid-based fire retardant. Construction and Building Materials. 2014; 72(15):1–6. DOI: 10.1016/j.conbuildmat.2014.09.004

[7] Giudice C, Alfieri P, Canosa G. Siloxanes synthesized "in situ" by sol-gel process for fire control in wood of *Araucaria angustifolia*. Fire Safety Journal. 2013; 61:348–354. DOI: 10.1016/j.firesaf.2013.09.013

[8] Double G, Urbanovich I, Zhurins A, et al. Application of analytical pyrolysis for wood fire protection control. Journal of Analytical and Applied Pyrolysis. 2007; 79(1-2):47–51. DOI: 10.1016/j.jaap.2006.10.019

[9] Chuang C, Tsai K, Yang T, et al. Effects of adding organo-clays for acrylic-based intumescent coating on fire-retardancy of painted thin plywood. Applied Clay Science. 2011; 53(4):709–715. DOI: 10.1016/j.clay.2011.06.009

[10] Bai G, Guo C, Li L. Synergistic effect of intumescent flame retardant and expandable graphite on mechanical and flame-retardant properties of wood flour-polypropylene composites. Construction and Building Materials. 2014; 50:148–153. DOI: 10.1016/j.conbuildmat.2013.09.028

[11] Baysal E, Altinok M, Colak M, et al. Fire resistance of Douglas fir (Pseudotsuga menzieesi) treated with borates and natural extractives. Bioresource Technology. 2007; 98(5):1101–1105. DOI: 10.1016/j.biortech.2006.04.023

[12] Östman B, Voss A, Hughes A, et al. Durability of fire retardant treated wood at humid and exterior conditions. Review of literature. Fire and Materials. 2001; 25(3):95–104. DOI: 10.1002/fam.758

[13] Zhou L, Guo C, Li L. Influence of ammonium polyphosphate modified with 3-(methylacryloxyl) propyltrimethoxy silane on mechanical and thermal properties of wood flour-polypropylene composites. Journal of Applied Polymer Science. 2011; 122(2):849–855. DOI: 10.1002/app.34069

[14] Gu J, Zhang G, Dong S, et al. Study on preparation and fire-retardant mechanism analysis of intumescent flame-retardant coatings. Surface and Coatings Technology. 2007; 201(18):7835–7841. DOI: 10.1016/j.surfcoat.2007.03.020

[15] Li G, Han J, Lou G, et al. Predicting intumescent coating protected steel temperature in fire using constant thermal conductivity. Thin-Walled Structures. 2016; 98(A):177–184. DOI: 10.1016/j.tws.2015.03.008

[16] Han Z, Fina A, Malucelli G. Thermal shielding performances of nano-structured intumescent coatings containing organo-modified layered double hydroxides. Progress in Organic Coatings. 2015; 78:504–510. DOI: 10.1016/j.porgcoat.2014.06.011

[17] Cavdar A, Mengeloğlu F, Karakus K. Effect of boric acid and borax on mechanical, fire and thermal properties of wood flour filled high density polyethylene composites. Measurement. 2015; 60:6–12. DOI: 10.1016/j.measurement.2014.09.078

[18] Keskin H, Atar M, Izciler M. Impacts of impregnation chemicals on combustion properties of the laminated wood materials produced combination of beech and poplar

veneers. Construction and Building Materials. 2009; 23(2):634–643. DOI: 10.1016/j.conbuildmat.2008.02.006

[19] Canosa G, Alfieri P, Giudice C. Nano lithium silicates as flame-retardant impregnants for *Pinus radiata*. Journal of Fire Sciences. 2011; 29(5):431–441. DOI: 10.1177/0734904111404652

[20] Pereyra A, Giudice C. Flame-retardant impregnants for woods based on alkaline silicates. Fire Safety Journal. 2009; 44:497–503. DOI: 10.1016/j.firesaf.2008.10.004

[21] Giudice C, Pereyra A. Silica nanoparticles in high silica/alkali molar ratio solutions as fire retardant impregnants for woods. Fire and Materials. 2010; 34:177–187. DOI: 10.1002/fam.1018

[22] Giudice C, Benítez J, Tonello M. Zinc borate and alumina trihydrate in chlorinated alkyd flame retardant coatings. Pitture e Vernici European Coatings. 2000; 76(10):17–24.

[23] Giudice C, Benítez J. Zinc borates as flame retardant pigments in chlorine-containing paints. Progress in Organic Coatings. 2001; 42:82–88. DOI: 10.1016/S0300-9440(01)00159-X

[24] Han Z, Fina A, Camino G. Organosilicon compounds as polymer fire retardants. In: Papaspyrides C and Kiliaris P, editors. Polymer Green Flame Retardants; 2014. pp. 389–418. DOI: 10.1016/B978-0-444-53808-6.00012-3

[25] Bardon J, Apaydin K, Laachachi A, et al. Characterization of a plasma polymer coating from an organophosphorus silane deposited at atmospheric pressure for fire-retardant purposes. Progress in Organic Coatings. 2015; 88:39–47. DOI: 10.1016/j.porgcoat.2015.06.005

[26] Meng X, Ragauskas A. Recent advances in understanding the role of cellulose accessibility in enzymatic hydrolysis of lignocellulosic substrates. Current Opinion in Biotechnology. 2014; 27:150–158. DOI: 10.1016/j.copbio.2014.01.014

[27] Kamide K. Cellulose in aqueous sodium hydroxide. In: Cellulose and cellulose derivatives. Molecular Characterization and its Applications, Elsevier B.V.; 2005. pp. 445–548. DOI: 10.1016/B978-044482254-3/50006-0

[28] Christodoulou C, Goodier C, Austin S, et al. Glass long-term performance of surface impregnation of reinforced concrete structures with silane. Construction and Building Materials. 2013; 48:708–716. DOI: 10.1016/j.conbuildmat.2013.07.038

[29] MacMullen J, Radulovic J, Zhang Z, et al. Masonry remediation and protection by aqueous silane/siloxane macroemulsions incorporating colloidal titanium dioxide and zinc oxide nanoparticulates: mechanisms, performance and benefits. Construction and Building Materials. 2013; 49:93–100. DOI: 10.1016/j.conbuildmat.2013.08.019

[30] Bücker M, Jäger C, Pfeifer D, et al. Evidence of Si-O-C bonds in cellulosic materials modified by sol-gel-derived silica. Wood Science and Technology. 2014; 48(5):1033–1047. DOI: 10.1007/s00226-014-0657-9

[31] Donath S, Militz H, Mai C. Wood modification with alkoxysilanes. Wood Science and Technology. 2014; 38:555–566. DOI: 10.1007/s00226-004-0257-1

[32] Li H, Hu Z, Zhang S, et al. Effects of titanium dioxide on the flammability and char formation of water-based coatings containing intumescent flame retardants. Progress in Organic Coatings. 2015; 78:318–324. DOI: 10.1016/j.porgcoat.2014.08.003

[33] Dasari A, Yu Z, Cai G, et al. Recent developments in the fire retardancy of polymeric materials. Progress in Polymer Science. 2013; 38(9):1357–1387. DOI: 10.1016/j.progpolymsci.2013.06.006

[34] Pereyra A, Canosa G, Giudice C. Nanostructured protective coating systems, fireproof and environmentally friendly, suitable for the protection of metallic substrates. Industrial & Engineering Chemistry Research. 2010; 49(6):2740–2746. DOI: 10.1021/ie901404s

Permissions

All chapters in this book were first published in WCE, by InTech Open; hereby published with permission under the Creative Commons Attribution License or equivalent. Every chapter published in this book has been scrutinized by our experts. Their significance has been extensively debated. The topics covered herein carry significant findings which will fuel the growth of the discipline. They may even be implemented as practical applications or may be referred to as a beginning point for another development.

The contributors of this book come from diverse backgrounds, making this book a truly international effort. This book will bring forth new frontiers with its revolutionizing research information and detailed analysis of the nascent developments around the world.

We would like to thank all the contributing authors for lending their expertise to make the book truly unique. They have played a crucial role in the development of this book. Without their invaluable contributions this book wouldn't have been possible. They have made vital efforts to compile up to date information on the varied aspects of this subject to make this book a valuable addition to the collection of many professionals and students.

This book was conceptualized with the vision of imparting up-to-date information and advanced data in this field. To ensure the same, a matchless editorial board was set up. Every individual on the board went through rigorous rounds of assessment to prove their worth. After which they invested a large part of their time researching and compiling the most relevant data for our readers.

The editorial board has been involved in producing this book since its inception. They have spent rigorous hours researching and exploring the diverse topics which have resulted in the successful publishing of this book. They have passed on their knowledge of decades through this book. To expedite this challenging task, the publisher supported the team at every step. A small team of assistant editors was also appointed to further simplify the editing procedure and attain best results for the readers.

Apart from the editorial board, the designing team has also invested a significant amount of their time in understanding the subject and creating the most relevant covers. They scrutinized every image to scout for the most suitable representation of the subject and create an appropriate cover for the book.

The publishing team has been an ardent support to the editorial, designing and production team. Their endless efforts to recruit the best for this project, has resulted in the accomplishment of this book. They are a veteran in the field of academics and their pool of knowledge is as vast as their experience in printing. Their expertise and guidance has proved useful at every step. Their uncompromising quality standards have made this book an exceptional effort. Their encouragement from time to time has been an inspiration for everyone.

The publisher and the editorial board hope that this book will prove to be a valuable piece of knowledge for researchers, students, practitioners and scholars across the globe.

List of Contributors

Adam Krajewski
Department of Wood Sciences and Wood Protection, University of Life Sciences - Szkoła Główna Gospodarstwa Wiejskiego w Warszawie, Nowoursynowska, Warszawa

Q. Angulo Ibáñez
Department of Architectural Constructions, Polytechnic University of Valencia, Valencia, Spain

Maria Cristina Porcu
University of Cagliari, Cagliari, Italy

Anil Akdogan and Ali Serdar Vanli
Mechanical Engineering Department, Yildiz Technical University, İstanbul, Turkey

Richard Hrčka and Marián Babiak
Department of Wood Science at Technical University in Zvolen, Zvolen, Slovak Republic

Ze-li Que, Zhe-rui Li, Xiao-lan Zhang and Biao Pan
College of Materials Science and Engineering, Nanjing Forestry University, Nanjing, China

Zi-ye Yuan
School of Foreign Studies, Nanjing Forestry University, Nanjing, China

Trulli Nicoletta, Monica Valdés and Barbara De Nicolo
Department of Civil Engineering, Environmental and Architecture, University of Cagliari, Cagliari, Italy

Massimo Fragiacomo
Department of Civil, Construction-Architecture and Environmental Engineering, University of L'Aquila, Italy

Valeria Awad, Linda Giresini, Mario Lucio Puppio and Mauro Sassu
Department of Energy, Systems, Territory and Constructions Engineering (DESTEC), University of Pisa, Pisa, Italy

Mikio Koshihara
Department of Human and Social Systems, International Center for Urban Safety Engineering (ICUS), University of Tokyo, Tokyo, Japan

Vincenzo De Luca
Dipartimento DIS, Università degli Studi della Basilicata, Potenza, Italy

Mojgan Nejad
Michigan State University, East Lansing, Michigan, USA

Paul Cooper
University of Toronto, Toronto, Canada

Carlos A. Giudice
UTN-FRLP (National Technological University-Faculty Regional La Plata), La Plata, Argentine

Guadalupe Canosa
UTN-FRLP (National Technological University-Faculty Regional La Plata), La Plata, Argentine
CIDEPINT (Research and Development Center in Coatings Technology), La Plata, Argentine

Index

A

Anatomical Direction, 81, 83, 87-89, 92-93, 96, 98

Anti-seismic Performance, 100, 113, 115, 117

C

Cellulose Fibers, 61, 63, 70

Compression Molding, 60, 70-71

Cross-laminated Timber, 57, 126, 141, 146-147, 163

Curvature Ductility, 36

D

Damage Rate, 29, 31

Degree Of Polymerization, 62-63

Dissipative Behavior, 35-36, 40, 42-45

Ductile Behavior, 35-36, 42-43, 45, 47-51

Dynamic Loads, 35, 45, 51-52, 166

E

Elastic Range, 35-36, 47

Electric Intensity, 13-14

Electric Resistance, 6

Equilibrium Temperature, 81, 96

Extrusion, 60-62, 66, 70-71, 73, 76-80

F

Fiber Load, 61

Fir Wood, 6, 8, 81, 90, 92-93

G

Gilded Timber, 1

Grain Deviation, 125

Gravitational Loads, 25

Gravity Loads, 26, 31

H

Heating Chamber, 12

I

Impact Hardness, 108-109

Infill Masonry, 38

L

Lignocellulosic Materials, 59, 223

Load-displacement, 36, 43-44, 51, 117, 158, 181

M

Machine Grading, 125-127, 136, 144-145

Magnetrons, 14

Matrix Polymers, 59-60

Microwave Chamber, 1, 5, 10

Microwave Radiation, 5-6, 13-16

O

Old House Borer, 1-10, 12-17

Oleophilic Compounds, 63

P

Pine Wood, 6, 8-11, 14-15, 69, 142, 196, 211

Plastic Matrix, 61, 75

Polymer Composites, 59-61, 73, 76-77, 79

Polystyrene, 63, 67

Pyrolysis, 9, 223-224

Q

Quasistationary Method, 81, 85-89

R

Radial Compress Strength, 108-109

Radiating Antenna, 6, 8-10, 14, 16

Resistance Sensor, 6

Round Pole Timber, 20, 22, 31-33

S

Sapwood Blocks, 6

Saw Mills, 60

Scanning Electron Microscope, 61, 203

Scattered Field, 1-2, 5, 14

Scattered Microwave, 10

Seismic Activity, 21, 32

Seismic Codes, 35, 49

Seismic Performance, 35, 49-50, 56-57, 100-101, 113, 115, 117, 121, 148

Seismic Stress Control, 36, 53

Shear Forces, 21, 66, 70

Short Radio Waves, 2

Specific Heat, 81-84, 87-88, 92-95, 99

Spruce Wood, 6, 8, 97, 181-182

Structural Strength, 21, 31

T

Tensile Forces, 22, 31

Tensile Resistance, 20

Thermal Conductivity, 81-84, 92-95, 98-99, 212, 224

Thermal Diffusivity, 81-82, 84, 87, 93-94, 96

Thermal Expansion, 81-82, 95-96

Thermic Conditions, 1-2

Thermic Numbness, 12

Thermoplastic Polymers, 62

Thermosetting Polymers, 62

Timber Buildings, 35-37, 40, 45, 47, 49, 53-54, 58, 101, 107, 123-124, 148

Timber Connections, 47-49, 51, 56

Timber Structures, 35-37, 39-40, 45-47, 49-51, 53, 55, 123-125, 142-145, 148, 163-166, 189-190

Timber-framed Buildings, 38-42

V

Vernacular Construction, 20

Vernacular Guidelines, 21

Vertical Cantilevers, 21, 29, 31

Visual Strength Grading, 126-127, 129, 132, 135, 137-138, 143-144

W

Wood Boring Insects, 1

Wood Disinfestation, 2

Wood Equilibrate, 82, 96

Wood Fiber, 59-61, 63-64, 66, 69, 75-76, 78-79, 99

Wood Filler, 59-66, 69, 71

Wood Powder, 59-61, 63, 66, 68-69, 72-73, 75, 78

Wood Reinforcement, 72

Wood Resin, 1

Wood Thermal Properties, 81-82, 89, 96-98

Wood-polymer Composite, 59

Wooden Bracing, 20-21, 23

Wooden Reinforcement, 20, 22-27, 29-30, 34

www.ingramcontent.com/pod-product-compliance
Lightning Source LLC
LaVergne TN
LVHW080712210926
848485LV00003B/455